Fiber Optic Pulse Compression

Numerical techniques and applications with MATLAB®

IOP Series in Advances in Optics, Photonics and Optoelectronics

SERIES EDITOR

Professor Rajpal S Sirohi Consultant Scientist

About the Editor

Rajpal S Sirohi is currently working as a faculty member in the Department of Physics, Alabama A&M University, Huntsville, Alabama (USA). Prior to this, he was a consultant scientist at the Indian Institute of Science, Bangalore, and before that he held a chair and was professor in the Department of Physics, Tezpur University, Assam. During 2000–11, he was an academic administrator, being vice-chancellor to a couple of universities and the director of the Indian Institute of Technology, Delhi. He is the recipient of many international and national awards and the author of more than 400 papers. Dr Sirohi is involved with research concerning optical metrology, optical instrumentation, holography, and speckle phenomena.

About the series

Optics, photonics, and optoelectronics are enabling technologies in many branches of science, engineering, medicine, and agriculture. These technologies have reshaped our outlook and our ways of interacting with each other and have brought people closer. They help us to understand many phenomena better and provide a deeper insight into the functioning of nature. Furthermore, these technologies themselves are evolving at a rapid rate. Their applications encompass very large spatial scales from nanometers to astronomical and a very large temporal range from picoseconds to billions of years. This series on advances on optics, photonics, and optoelectronics aims to cover topics that are of interest to both academia and industry. Some of the topics to be covered by the books in this series include biophotonics and medical imaging, devices, electromagnetics, fiber optics, information storage, instrumentation, light sources, CCD and CMOS imagers, metamaterials, optical metrology, optical networks, photovoltaics, free-form optics and its evaluation, singular optics, cryptography, and sensors.

About IOP ebooks

The authors are encouraged to take advantage of the features made possible by electronic publication to enhance the reader experience through the use of color, animation, and video and by incorporating supplementary files in their work.

Do you have an idea for a book you'd like to explore?

For further information and details of submitting book proposals, see iopscience.org/books or contact Ashley Gasque on Ashley.gasque@iop.org.

A full list of titles published in this series can be found here: https://iopscience.iop.org/bookListInfo/series-on-advances-in-optics-photonics-and-optoelectronics.

Fiber Optic Pulse Compression

Numerical techniques and applications with MATLAB®

R Vasantha Jayakantha Raja and A Esther Lidiya

Department of Physics, School of Electrical and Electronics Engineering, SASTRA Deemed to be University, Thanjavur, Tamil Nadu, India

IOP Publishing, Bristol, UK

ISBN 978-0-7503-2686-5 (ebook)
ISBN 978-0-7503-2684-1 (print)
ISBN 978-0-7503-2687-2 (myPrint)
ISBN 978-0-7503-2685-8 (mobi)

DOI 10.1088/978-0-7503-2686-5

Version: 20220801

IOP ebooks

British Library Cataloguing-in-Publication Data: A catalogue record for this book is available from the British Library.

Published by IOP Publishing, wholly owned by The Institute of Physics, London

IOP Publishing, Temple Circus, Temple Way, Bristol, BS1 6HG, UK

US Office: IOP Publishing, Inc., 190 North Independence Mall West, Suite 601, Philadelphia, PA 19106, USA

To my better half dhayasagari who encouraged me to go on every adventure and my daughter thanvika hasini who makes me so proud.

—Vasanth

Contents

Preface

The generation of ultrashort pulses in the visible and near-infrared regions of the electromagnetic spectrum is extremely useful in many fields, including ultrafast spectroscopy, nanosurgery, micromachining, extremely nonlinear optics, and terahertz pumping sources. In general, the generation of shorter pulses requires more complex and expensive state-of-the-art setups than those required to generate longer pulses. However, the pulse-compression process has been discovered to be one of the best techniques for obtaining ultrashort pulses via optical fibers. Mollenauer *et al* discovered the use of pulse compression in fiber to generate ultrashort pulses. Since then, it has gained traction in a variety of nonlinear optics fields. Light manipulation became much easier after the invention of photonic crystal fiber (PCF) due to its unique properties such as its choice of dispersion properties, endless single-mode operation, high nonlinearity, large numerical aperture, and so on. These PCF properties pave the way for new opportunities in the field of pulse compression. Recently, it has become possible to generate desired pulse widths of ultrashort pulses using a pulse-compression technique in PCF, whereby anomalous dispersion can be obtained from the visible range to IR. A variety of theoretical and experimental approaches that use PCFs have already been proposed.

The primary goal of this book is to discuss the theoretical conditions required to generate pedestal-free ultrashort pulses with improved properties in PCF, as well as to investigate the nonlinear phenomena of ultrashort pulses with various compression schemes. To calculate fiber parameters that depend on the physical environment, such as dispersion, nonlinearity, and loss, the PCF is numerically modeled using the fully vectorial effective index method (FVEIM). To generate ultrashort pulses, the calculated environment-dependent fiber parameters are used to numerically solve the modified nonlinear Schrödinger equation (MNLSE). The MNLSE takes into account the influences of dispersion and nonlinear parameters in order to produce high-quality ultrashort pulses. In addition, the pulse's quality is numerically calculated by investigating the compression factor, quality factor, and pedestal energy. The proposed compressors are also optimized by varying the fiber structure, pump wavelength, and pump power.

This book explains the generation of ultrashort pulses in PCF using different pulse-compression techniques, such as higher-order soliton compression, adiabatic compression, self-similar compression, cascaded compression, and supercontinuum generation (SCG)-induced compression. The use of pulse-compression techniques varies, depending on the application requirements, and each compression scheme has its own set of benefits and drawbacks. The nonlinear optical dynamics of such compression schemes are thoroughly investigated in the context of the use of PCF. Initially, the two most commonly used compression techniques, higher-order soliton compression and adiabatic compression, are discussed. Although the former can be compressed to a high degree, the compressed pulses suffer from significant pedestal generation, resulting in nonlinear interactions between neighboring solitons. A dispersion map with monotonically decreasing dispersion along the propagation direction is typically used in adiabatic soliton compression. The soliton self-adjusts to maintain the balance between dispersion and nonlinearity if the dispersion varies slowly enough.

The following chapter focuses on proposing a compact compressor in the near-visible regime using a novel compression technique based on self-similar analysis. Using a suitable PCF design and this self-similar based compression technique, pulse compression with a high compression factor and low pedestal energy can be obtained over a short distance with a low input power. Following that, a cascaded compression scheme for generating ultrashort pulses with a temperature-controlled chloroform-infiltrated PCF is proposed. The majority of the previously proposed cascading higher-order soliton compression schemes work by selecting different fiber couplings and optical elements to generate a compressed pulse from compressed input pulses. Rather than using multiple fibers with varying dispersion and nonlinear parameters, we use only one fiber that is exposed to three different temperatures. Finally, we discuss pulse compression accomplished through the use of SCG, which is linked with an investigation of ultrashort pulses and broadband spectra in PCF. The supercontinuum (SC) process produces intense ultrafast broadband highly coherent pulses spanning a few octaves; it has emerged as the technology of choice for the future generation of broadband sources and pulse compression and has been found to be one of the best techniques for obtaining ultrashort pulses via optical fibers. The soliton fission technique will be used to investigate the impact of temperature in the SCG process to support the physical explanation of tunable ultrashort pulses.

Acknowledgements

There are numerous people who have helped us to complete this work. We were very fortunate to have the opportunity to interact with outstanding researchers, teachers, colleagues and students, and we profited greatly from discussions with them to write this book. It gives us immense pleasure to express our deep sense of gratitude to all who offered numerous suggestions and many constructive criticisms in this exciting field of 'pulse compression'. First, We would like to specifically thank Professor K Porsezian (Pondicherry University) for his genuine interest in our scientific progress and many of his illuminating discussions on nonlinear pulse propagation in photonic crystal fiber (PCF) for various problems. Then, we wish to express appreciation to Professor Anton Husakou (Max-Born Institute, Germany) with whom we collaborated in the study of higher order solitons; his inextinguishable enthusiasm for numerical simulations and willingness to share his knowledge is an important aspect in our book. We are truly thankful to Professor K Senthilnathan (VIT University) for his discussions and help with our work on pulse compression. In particular, his help in understanding how self-similar technique is implemented in pulse compression is appreciated. We also thank Professor Shailendra K Varsheney (IIT Kharaghpur) and Professor Balaji Srinivasan (IIT Madras), for their tireless help whenever we had PCF-related questions. We would also like to thank Professor K Nakkeeran (University of Aberdeen) for his assistance with the implementation of the various algorithms to make this book possible. Finally, we are indebted to all the experts who helped us directly or indirectly in preparing this book.

Author biographies

R Vasantha Jayakantha Raja

R Vasantha Jayakantha Raja is currently an assistant professor at the Department of Physics, SASTRA Deemed to be University, Thanjavur, India. He is an expert in theoretical and numerical modeling of nonlinear pulse propagation with vast experience in realistic numerical modeling of experiments. He has been actively working in the research field of nonlinear fiber optics and ultra-fast optics. He has mainly focused his research on numerical modeling of nonlinear pulse propagation through photonic crystal fiber (PCF) for various nonlinear applications, including supercontinuum generation (SCG) for broadband light sources, generating ultrashort pulses using pulse compression, frequency conversion through modulational instability and four-wave mixing process. His current research concern is ultra-fast femtosecond optics and optical parametric amplification based on four-wave mixing process. He successfully completed a research project under the Young Scientist Scheme by the Department of Science and Technology, extra-mural research fund scheme from CSIR and ASEAN-India collaborative research project under AISTDF-DST scheme. Right now, he is sanctioned by DST through Indo-Uzpekistan collaborative research scheme. He has supervised two PhD students and is currently supervising another PhD student. He had already authored one book on SCG and two book chapters, including nonlinear pulse propagation in PCFs, and on general nonlinear effects. He has authored and coauthored more than 35 research papers in peer-reviewed journals and more than 55 papers in conference proceedings. Also, he is a member of the Optical Society of America (OSA).

A Esther Lidiya

A Esther Lidiya received her BSc, MSc, and MPhil degrees in Physics from the Gandhigram Rural Institute Deemed University, Dindigul, Tamil Nadu, India. She has submitted her PhD thesis at the Department of Physics, SASTRA Deemed to be University, Thanjavur, Tamil Nadu, India. Her research interests include the modeling of PCF-based pulse-compression systems for the generation of high-power ultrashort pulses, fiber lasers/amplifiers, and surface plasmon resonance sensors.

IOP Publishing

Fiber Optic Pulse Compression
Numerical techniques and applications with MATLAB®
R Vasantha Jayakantha Raja and A Esther Lidiya

Chapter 1

Introduction

This chapter is intended to provide a brief introduction to ultrashort pulses and an overview of the evolution of the existing ultrashort pulse-generation techniques. The characteristics of ultrashort pulses that are significant for its application in diverse fields of science and technology are elaborated.

1.1 Ultrashort pulses

Science and technology need continuous revolutions, renovations, and replacements; for example, the invention of light amplification by stimulated emission of radiation (LASER) in 1960 revolutionized the science and engineering sectors by enabling the generation of highly coherent, monochromatic optical pulses [1], which acted as a gateway to novel spectroscopic methods of investigation that deepened our understanding of atomic structure. The time resolution of the equipment at our disposal limits our capacity to detect natural dynamics that occurs in short time intervals. For example, mechanical shutters provide millisecond resolution, stroboscopic lighting allows us to explore the microsecond range, and modern electronic sampling oscilloscopes have reduced the limit to the picosecond range. Beyond this limit, the development of ultrafast lasers has advanced the temporal resolution of measurement by another three orders of magnitude into the sub-ten-femtosecond regime, allowing for the direct observation of vibrational molecular dynamics [2]. In general, electromagnetic pulses of optical energy that have pulse durations of a picosecond (10^{-12} s) or less are called ultrashort pulses. The word 'ultra,' meaning 'beyond,' originates from the Latin word 'ulter.' The first ultrashort pulse laser was demonstrated by De Maria *et al* after six years of the invention of Maiman's laser, using a passively mode-locked Nd:glass laser with an approximately measured minimum pulse width of $\approx 3.7 \times 10^{-13}$ s [3]. The ultrashort pulses are highly collimated beams, hence they propagate in a well-defined direction. Because of the high spatial coherence of ultrashort pulses, they can focus on very small spots, as small as 1 μm^2. A small spot size combined with a short pulse duration results in extremely high optical intensity. The spatial extent of short temporal pulses helps to put them in perspective. For example, a one-second light pulse may transverse a distance of 300 000 km, which is equal to the speed of light multiplied by one second. Meanwhile, a picosecond pulse has a spatial extent of 0.3 mm and a femtosecond pulse has a spatial extent of 0.3 μm. As a result, ultrafast events can be determined by these short pulses. The temporal confinement of light to durations close to the optical period [4] and the conversion of pulses of few cycles to extreme ultraviolet and x-ray wavelengths [5, 6] have enabled the measurement and control of electron dynamics on the sub-femtosecond timescale [7, 8]. The ability to scale peak and average power is an essential characteristic of the development of femtosecond laser technology [9]. An astounding capability to enhance laser peak power has been achieved by various ultrashort pulse-amplification techniques, such as chirped-pulse amplification (CPA), optical parametric chirped-pulse amplification (OPCPA) [10], and backward Raman scattering (BRA) during the last 25 years, and it has revolutionized laser science. These techniques help to amplify ultrashort pulses up to the peak power of PW without damaging the optical medium.

Even though these ultrashort pulse-amplification techniques provide light sources capable of generating hundreds of watts of average power; they seldom generate pulses shorter than 100 fs, which are extremely useful for frequency conversion to the extreme ultraviolet or the mid-infrared region. As a result, the development of power-scalable pulse-compression techniques is currently a work in progress. For the high-quality generation of ultrashort pulses, existing laser sources such as mode-locked lasers and Mamyshev oscillators require external pulse-compression techniques. In this context, several pulse-compression approaches based on linear and nonlinear optical components have been developed in an attempt to generate optical pulses with durations ranging from tens to hundreds of femtoseconds throughout a wavelength band spanning from the ultraviolet to the far infrared [11]. The confinement of all optical energy to a short time interval provides access to unprecedented peak powers, resulting in predominantly nonlinear interactions between light and matter, which has led to tremendous evolution in the field of ultrafast optics and opened new directions in the frontiers of high-field science and ultrafast spectroscopy [12, 13]. This book deals with such nonlinear pulse-compression techniques used for the generation of high-quality ultrashort pulses.

1.2 Characteristics of optical pulses

Ultrashort pulses are electromagnetic wave packets characterized by a time- and space-dependent electric field, which are the measurable quantities that are directly connected with the electric field. A complex representation of the field amplitude is especially useful when dealing with pulse propagation. The propagation of such fields and their interaction with matter are regulated by Maxwell's equations in a semi-classical manner, and the material response is represented by a macroscopic polarization. In general, the complex electric field of an optical pulse $E(t)$ as it propagates down the z-axis is usually expressed in the time domain by an envelope which is the product of an amplitude function and a phase term, as follows:

$$E(z, t) = U(z, t)e^{i(\omega_0 t - k_0 z + \phi(t))}, \tag{1.1}$$

where $U(z, t)$ is the slowly varying envelope of the wave packet, $\phi(t)$ is the temporal phase, ω_0 is the carrier frequency, and $k_0 = \frac{n(\omega_0)\omega_0}{c}$ is a wave number that determines the carrier wavelength $\left(\text{as } \lambda_0 = \frac{2\pi}{k_0}\right)$ of the wave packet. The time-varying temporal phase function $\phi(t)$ establishes a time-dependent carrier frequency (instantaneous frequency) $\omega(t) = \omega_0 + d\phi(t)/dt$. The most convenient way to describe the wave number in a Taylor expansion around the carrier frequency is

$$\begin{aligned} k(\omega) = k(\omega_0) &+ \left.\frac{\partial k}{\partial \omega}\right|_{\omega_0} (\omega - \omega_0) + \frac{1}{2}\left.\frac{\partial^2 k}{\partial \omega^2}\right|_{\omega_0} (\omega - \omega_0)^2 \\ &+ \frac{1}{6}\left.\frac{\partial^3 k}{\partial \omega^3}\right|_{\omega_0} (\omega - \omega_0)^3 + \cdots. \end{aligned} \tag{1.2}$$

Figure 1.1 depicts an electric field that has a pulse duration of 10 fs and its corresponding intensity pulse envelope. One can also describe the optical pulse using the spectral domain obtained from the time domain, in which the pulse consists of different frequency components. In general, the optical pulse in the spectral domain is described by its amplitude $U(z, \omega)$ and spectral phase $\phi(z, \omega)$ as follows:

$$\tilde{E}(z, \omega) = U(z, \omega)e^{i\phi(z, \omega)} \tag{1.3}$$

Given the temporal dependence of the electric field $E(t)$, the complex spectrum of the field strength $\tilde{E}(\omega)$ can be derived mathematically through the complex Fourier transform ($\mathcal{F}$):

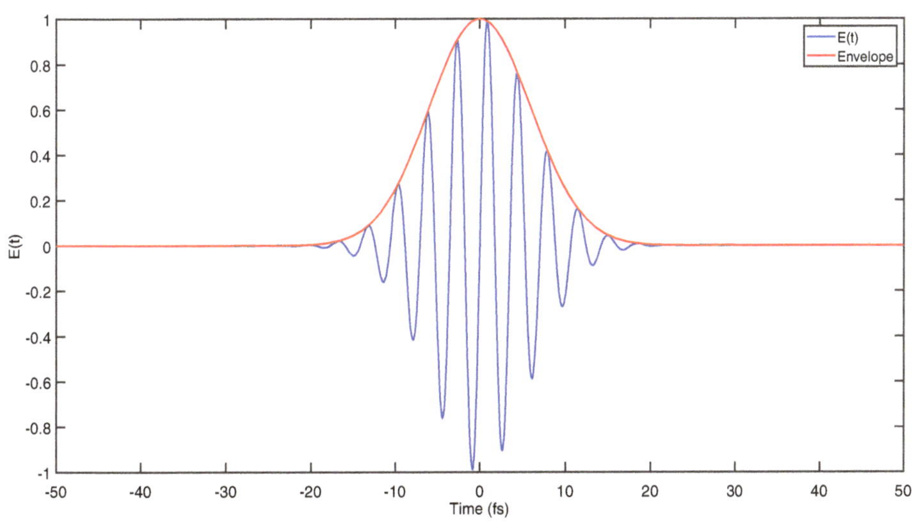

Figure 1.1. Electric field profile of an optical pulse. The red line indicates the corresponding intensity profile.

$$\tilde{E}(z, \omega) = \mathcal{F}[E(t)] = \int_{-\infty}^{\infty} E(t)e^{-i\omega t}dt. \quad (1.4)$$

Given $\tilde{E}(\omega)$, the time-dependent electric field can be obtained through the inverse Fourier transform ($\mathcal{F}^{-1}$):

$$E(t) = \mathcal{F}^{-1}[\tilde{E}(\omega)] = \frac{1}{2\pi}\int_{-\infty}^{\infty} \tilde{E}(\omega)e^{i\omega t}d\omega. \quad (1.5)$$

For an electric field $U(z, t)$ propagating in a dispersionless material of refractive index n, the optical pulse energy can be easily calculated if the pulse parameters such as the pulse width (τ), peak power (P_0) and repetition rate T_{rep} are known. Figure 1.2 provides a pictorial representation of the measurement of a laser pulse's peak power, pulse width, and repetition rate. The pulse width in the time domain that spans a pulse is commonly calculated using the pulse's full width at half maximum (FWHM). For example, the pulse width of a Gaussian pulse is related to FWHM as follows (figure 1.3):

$$T_{\text{FWHM}} = 2\sqrt{ln\ 2} \times \tau \approx 1.665\tau. \quad (1.6)$$

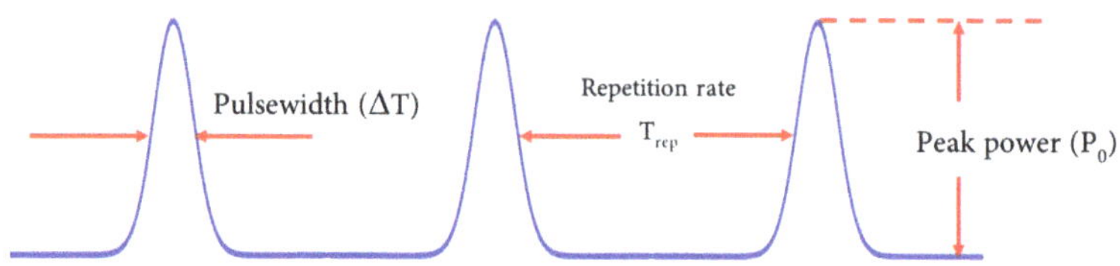

Figure 1.2. Measurement of the peak power, pulse width, and repetition rate.

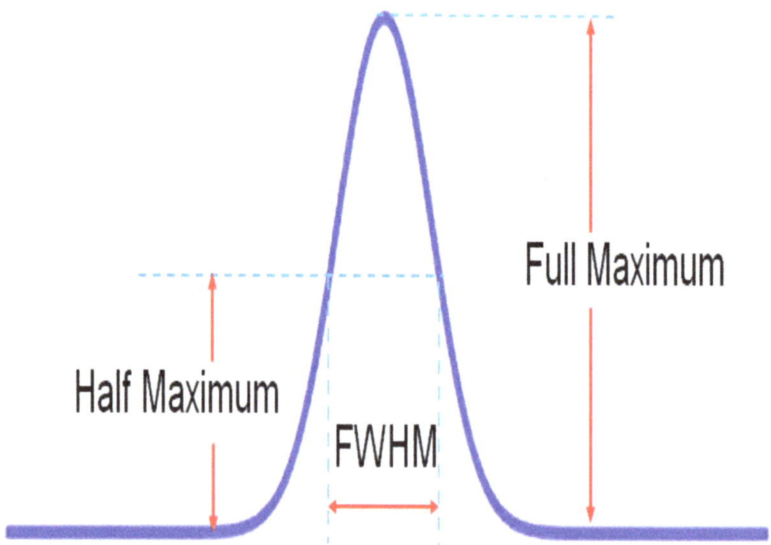

Figure 1.3. Representation of the FWHM in an optical pulse.

Given this value, the pulse energy can be calculated as follows:

$$E = \int_{-T_{\text{rep}}}^{T_{\text{rep}}} |U(z, t)|^2 dt, \tag{1.7}$$

where $P_0 = |U(z, t)|^2$. The energy per unit area is the instantaneous intensity (Wm^{-2}) of a field and can be calculated from

$$I(t) = E^*(t)E(t) = |E(t)|^2 \tag{1.8}$$

and the energy density per unit area (J cm^{-2}),

$$W = \int_{-\infty}^{\infty} I(t')dt'. \tag{1.9}$$

1.3 Generation of broadband spectra

Laser light pulses with time durations on the order of a picosecond or less have a broad spectrum, which can be observed in packets of waves that are extremely localized in time. In this section, we will study about how the wave packets of ultrashort pulses can create a broad spectrum. An ideal light wave has a single frequency and it is spatially unlocalized. Consider the propagation of one light wave along the x-axis, as shown in figure 1.4, in the presence of the simultaneous propagation of light waves of equal amplitude but slightly different frequency. When these pulses, which have alternate out-of-phase and in-phase relationships, are added together, the single light wave is divided into beats via superposition. Figure 1.4 depicts the results of superimposing three waves, five waves, and seven waves to generate a longer, single light pulse.

As in the generation of ultrashort pulses by combining a large number of slightly varying frequency components, as depicted in figure 1.4, by adding a larger number of slightly different frequency components, one can generate ultrashort pulses as shown in figure 1.5, which proves that an ultrashort pulse may be generated by broadening the spectrum of a longer pulse. However, due to the broad optical bandwidth of ultrashort pulses, chromatic aberrations that occur in the focusing optics can lead to complicated spatio-temporal effects, which may cause the focused pulse to have a larger duration than it had before focusing. To overcome this issue, refractive and diffractive optics with suitable lens combinations are required.

1.4 Time–bandwidth product

It is important to acknowledge the relationship between spectral width and pulse duration when considering the generation of ultrashort pulses. By measuring the pulse width in both the frequency and time domains, we can compare the results to the Heisenberg uncertainty principle, which limits the minimum uncertainty in these variables. The energy–time uncertainty principle is given by

$$\Delta E \Delta T \geqslant \frac{\hbar}{2}, \tag{1.10}$$

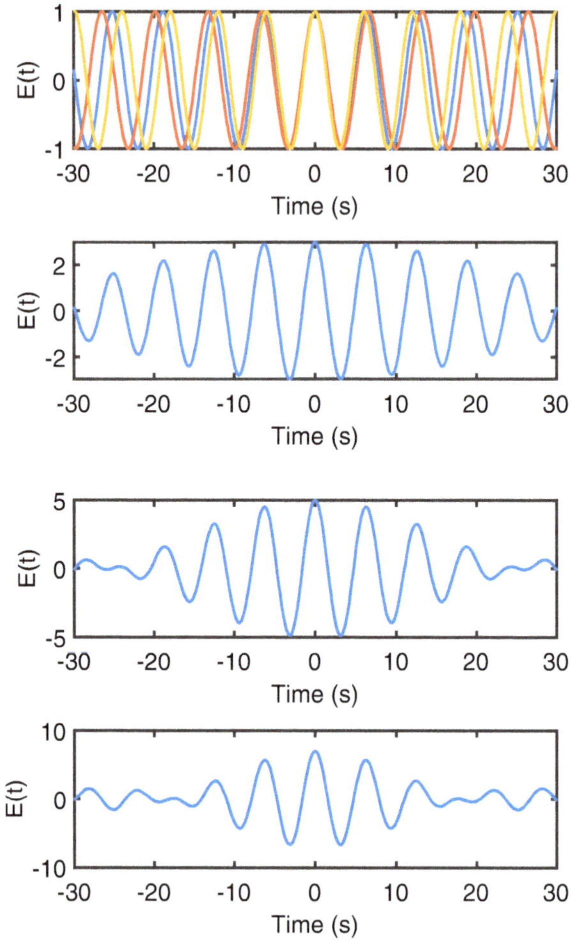

Figure 1.4. Superpositions of three waves, five waves, and seven waves.

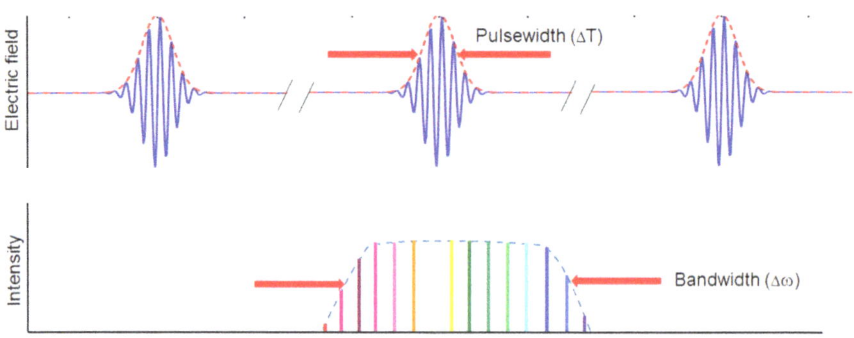

Figure 1.5. Electric field profile of an ultrashort pulse at a central wavelength of λ_0. The dashed line indicates the corresponding temporal and spectral intensity profile.

where ΔE is the standard deviation in the energy and ΔT represents the deviation in time. Since, for photons, $E = \hbar\omega$, equation (1.10) can be written as

$$\Delta T \Delta\omega \geqslant \frac{1}{2}. \tag{1.11}$$

It can be seen that a smaller ΔT demands a larger $\Delta\omega$ or frequency range. The product of pulse duration and spectral bandwidth is known as the time–bandwidth product (TBP).

In theory, this indicates that a broad spectral bandwidth ($\Delta\omega$) is required to generate a short pulse of light with a specific duration (ΔT). According to figure 1.6, if one chooses a large bandwidth, the pulse width is reduced. The reverse is also true: a pulse's TBP is always greater than the theoretical minimum given by the uncertainty principle (for the appropriate width definition). When equality to 1/2 is reached in the context of equation (1.11), the pulse involved is called a Fourier-transform-limited pulse. The variation in the phase of such a pulse is beautifully uniform; thus, it has a linear time dependence and its instantaneous frequency becomes time independent. In real-time measurements, the TBP is a measure of the complexity of a wave or pulse. The TBPs of actual signals vary, but there is always a minimum TBP for a certain desired effect. In communications over a channel, transmitting a certain amount of data over a given bandwidth requires a certain time. The TBP measures how well one can use the available bandwidth for a given channel. Even though every pulse's time-domain and frequency-domain functions are related by the Fourier transform, a wave with the minimum TBP is called Fourier transform limited. For ideal pulses, the product of the pulse width multiplied by the bandwidth has a minimum constant value. More commonly, the pulse duration is defined according to the principle of the FWHM of the optical power versus time. Equation (1.11) then becomes

$$\Delta\nu\Delta t \geqslant K, \tag{1.12}$$

where $\Delta\nu$ is the frequency at the FWHM and Δt is the duration at half maximum. The value of K depends upon the symmetrical shape of the pulse. For a Gaussian function, $K = 0.441$, and for a hyperbolic secant function, $K = 0.315$.

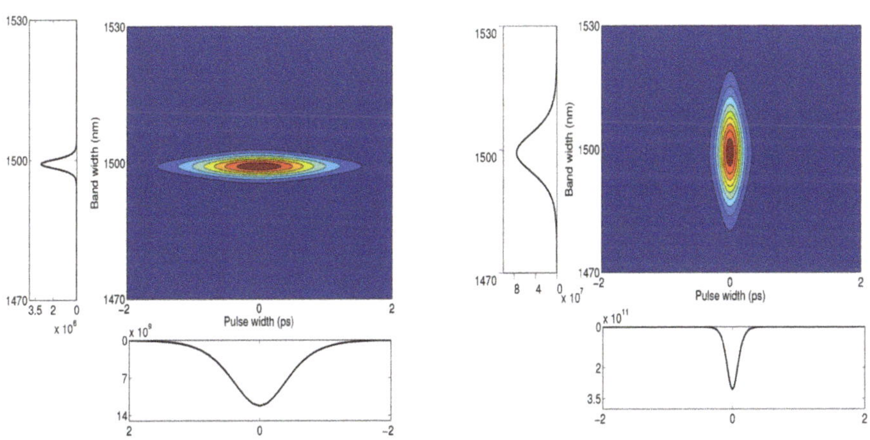

Figure 1.6. Simulated plotted spectrograms for (a) a 1 ps pulse and (b) a 200 fs pulse.

1.5 Applications of ultrashort pulses

The applications of ultrashort pulses stem from direct laser beam interactions with matter and from the interactions between matter and the secondary particle and photon sources they drive. Ultrashort pulses are helpful in studies of the first principles of fundamental processes and to monitor extremely fast events in biology (photosynthesis, vision, protein folding), chemistry (molecular vibrations, re-orientations, and liquid-phase collisions), and electronic processes (high-lying excited-state lifetimes, photo-ionization, and electron–hole relaxation times that determine the response times of light detectors and electronics). Because of the broad spectrum of ultrashort pulses, they are used in medical diagnostics, such as optical coherence tomography (OCT), hard tissue ablation, and brain surgery, frequency metrology [14, 15], standoff trace gas detection (particularly in the oil and gas industry), and the detection of chemical components in artifacts [16–19]. The extreme concentration of energy in femtosecond pulses is useful for the materials processing of semiconductors, composite materials, glasses, and plastics (because much finer structures can be created in the absence of thermal interaction caused by longer pulses), and in attosecond pulse generation, laser-driven particle acceleration, and defense applications [20–22]. The latest advances in femtosecond technology have strongly emphasized the control of ultrashort pulses in many applications in which the preservation of the pulse duration is most important. Some of the applications of ultrashort sources are elaborated below.

1.5.1 Frequency metrology

Because of their broad optical bandwidth, ultrashort pulses are used in the precise measurement of optical frequency, for example in frequency metrology as in figure 1.7. Stabilized laser emission is known to be composed of various spectral lines known as a 'frequency comb,' which is a highly precise tool for the detection of different light frequencies. A frequency comb can serve as a 'ruler' in the measurement of unknown frequencies by allowing them interfere with the comb and measuring the beat frequency. However, the unknown phase shift between the

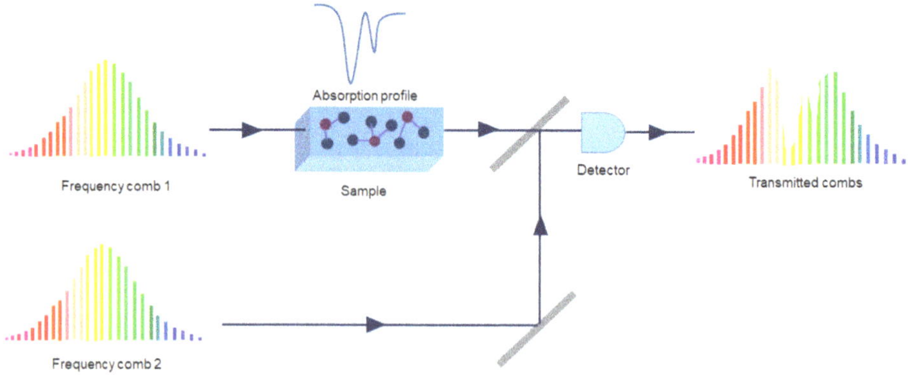

Figure 1.7. Application of ultrashort pulses in frequency metrology.

envelope and the carrier causes a drift of the comb, which is the fundamental issue with this approach. This is where ultrashort sources come into play: by comparing the frequencies of octave-separated comb peaks, the phase shift may be measured and controlled. A true optical-frequency comb contains very small, evenly spaced teeth that cover the whole visible spectrum of light. The teeth may be used as a ruler to precisely measure the light produced by lasers, atoms, stars, and other objects with extreme accuracy.

1.5.2 Optical coherence tomography

OCT is a micrometer-scale high-resolution cross-sectional imaging technology used to acquire pictures of strongly scattering media. Due to the interferometric basis of this technique, the resolution depth $\triangle z$ of the cross-sectional image is related to the central wavelength λ_c and the FWHM bandwidth $\triangle\lambda$ of the light source as follows (figure 1.8):

$$\triangle z = \frac{2\ln 2}{\pi}\frac{\lambda_c^2}{\triangle\lambda} \approx 0.44\frac{\lambda_c^2}{\triangle\lambda}. \tag{1.13}$$

If the source spectrum is approximately Gaussian in form, the optimal value of λ_c is determined by the medium being studied. To achieve a good penetration depth, the 800 nm wavelength region is optimal for OCT measurements of the eye, due to its lower absorption, whereas the 1300nm wavelength region is best for observations of highly scattering tissue, such as skin. Based on the foregoing, it is desirable to have

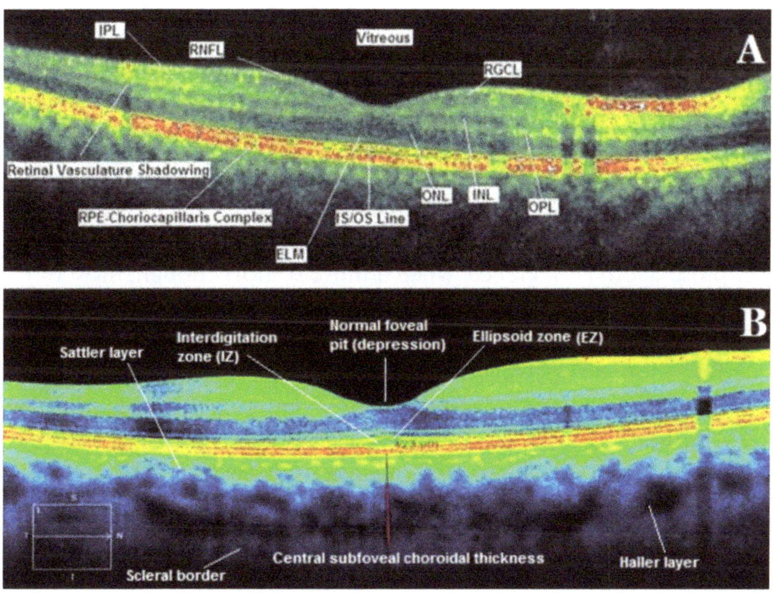

Figure 1.8. Spectral domain (a) and enhanced depth imaging (b) OCT scans of a healthy subject, showing the main retinal and choroidal histology landmarks. (Courtesy: Turgut *et al* [71]. Reproduced with permission.)

an OCT light source with an exceptionally broad (hundreds of nanometers), relatively smooth, and flat spectrum, with a central wavelength tailored to the specific OCT application.

1.5.3 Wavelength-division multiplexing

Ultrashort laser pulses have broad spectral bandwidth, which is essential to research into high-speed transmission systems with extremely high data rates, notably femtosecond wavelength-division multiplexing (WDM) [23]. WDM technology is a promising way to increase the capacity of backbone networks by providing low-noise multiple wavelength sources with very high wavelength accuracy, high amplitude stability, and high brightness. In WDM applications, the data rate is secondary to the spectral bandwidth of the femtosecond source. Standard WDM systems utilize a single-frequency laser for each WDM channel. The wavelengths of each laser need to be individually controlled and stabilized to ensure that they coincide with the preassigned WDM channels. In order to alleviate this problem, broadband sources that simultaneously cover all channels are needed. WDM systems with dense channel spacing are the best alternative for increasing capacity in light-wave transmission systems across both short and long distances. In this regard, femtosecond lasers are excellent diffraction-limited broadband sources for WDM that are more stable and reliable than the supercontinuums generated by nonlinear processes in fibers. A 100 fs laser pulse has a spectral bandwidth of approximately 3 THz, enough for 30 channels spaced at 100 GHz, a channel spacing that has been proposed as a standard [24].

1.5.4 Materials processing

Laser materials processing is a major component of manufacturing and is used to accomplish tasks ranging from heating for hardening, melting for welding and cladding, and the removal of material for drilling and cutting [25]. Material removal is based on the fact that all materials have an ablation threshold, i.e. a point at which they are directly vaporized when hit with a laser beam of sufficient peak optical intensity [26]. The threshold fluence (energy per unit area) for ablation reduces as a function of pulse width [27]. Hence, when a pulse is short enough, most of the optical pulse excites electrons, which then quickly cause a small section of the material to ablate without heating the substrate during the interaction. As a result, laser materials processing via plasma formation is considered to be a 'cold process' [28]. Because of the minimum thermal energy deposition in materials, ultrashort pulses allow for highly precise cutting, resulting in high-aspect-ratio holes and finely imprinted patterns with no collateral damage outside the desired interaction volume [29]. The resulting ejected material is mostly gaseous or very fine particles, and leaves behind a very limited heat-affected zone (HAZ), typically much less than a micron. Short pulses are also used in surface processing in order to clean or texture surfaces, resulting in hydrophobic surfaces or chemically reactive surfaces [30]. The typical intensities required for such tasks include heat treatment at 10^3–10^4 W cm^{-2}, welding and cladding at 10^5–10^6 W cm^{-2}, and material removal at 10^7–10^9 Wm^{-2} for

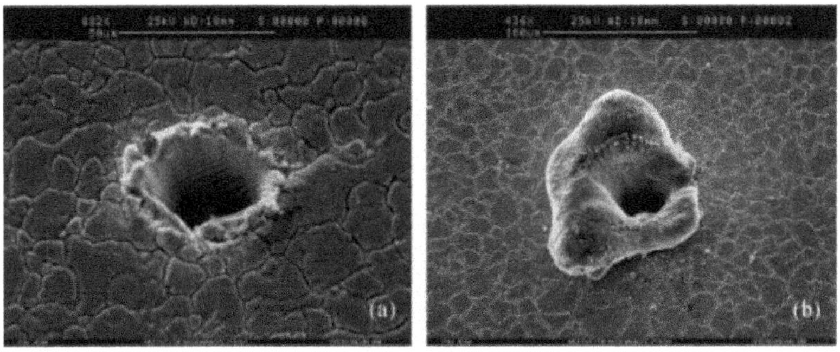

Figure 1.9. Scanning electron microscopy pictures of the entrances of holes drilled in 0.5 mm-thick stainless steel sheets using the percussion drilling technique with an energy of 30 μJ and repetition rates of (a) 100 kHz and (b) 400 kHz. (Courtesy: Ancona *et al* [72]. Reproduced with permission.)

drilling, cutting, and milling [31]. The typical operating parameters of the commercial lasers used for manufacturing include pulse widths of 100–200 fs, peak energies of 50–150 μJ, average powers of 100–150 W, and pulse repetition rates of up to 1 MHz. Due to the ability of femtosecond lasers to efficiently fabricate complex structures, state-of-art laser processing techniques with ultrashort pulses are used to structure materials with sub-micrometer resolution, such as intricate three-dimensional photonic crystals, micro-optical components, gratings, and optical waveguides. Such structures rely on the creation of increasingly sophisticated miniature parts. As a result of the precision, fabrication speed, and versatility of ultrafast laser processing, it is it well placed to become a vital industrial tool for advanced material 3D micro/nano processing (figure 1.9).

1.5.5 Medicine

In the field of medicine, lasers have reduced the need for sterilization or anesthetics. Ultrashort pulsed laser technology is now commonly used in the medical industry to fabricate high-quality surgical stents that have micron-scale features such as 1 μm-diameter holes with a large length-to-diameter ratio. Single-mode femtosecond laser technology is proving the best tool for these needs. The femtosecond laser ablation depths achieved using a single laser pulse can be more precise than those of material removed by conventional laser melting. Cracks due to thermal damage appear as a result of picosecond to femtosecond pulses but nearly disappear when the pulse duration is reduced to 5 fs. Because of the reduced collateral damage, high-intensity ultrashort pulses are used in various kinds of surgery based on laser processing of tissues in which intense laser pulses are delivered to internal tissues via optical fibers (figure 1.10).

Particularly well known is laser-assisted in-situ keratomileusis (LASIK), which uses ultrafast laser scalpels to make incisions in the eyeball as part of a laser sculpting protocol to improve eyesight [32]. In these surgeries, the peak power is limited by the microscopic nature of the instrumentation. In addition, the promised ability of ultrahigh-intensity laser pulses to create different kinds of high-energy

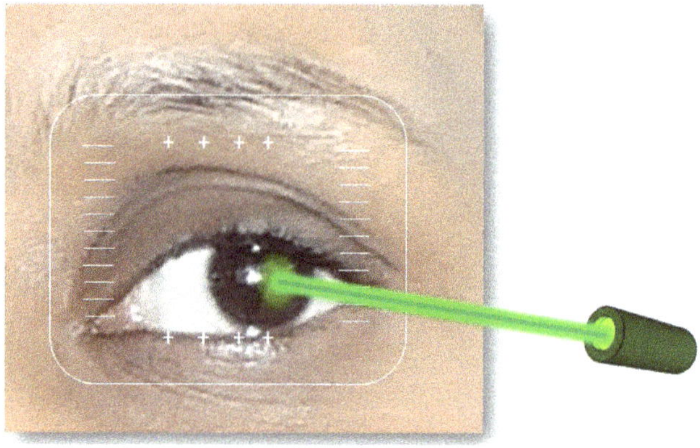

Figure 1.10. Laser eye surgery involves reshaping the cornea of the eye.

particles and radiation through interaction with a variety of sources leads to applications such as hard x-ray and γ-ray imaging which offer imaging features that cannot be achieved using ordinary x-rays [33], therapies using high-energy x-rays and γ-rays, therapies using laser-accelerated electron beams, therapies using laser-accelerated ion beams (mostly protons), and transmutation to create radioactive positron sources for positron emission tomography (PET) [34, 35]. Likewise, radiotherapy requires high-energy electrons to selectively kill cancer cells. In this regard, high-intensity ultrafast lasers based on wake field acceleration are commonly used instead of cyclotrons and linacs for the generation of high-energy electrons [36, 37]. Since thermal damage and stress-induced cracking depend on the average power of the femtosecond laser source, absorbed laser power leads to melting or thermal shock, even with picosecond or femtosecond pulse durations. Hence, low-average-power or ultrashort low-energy pulses are needed in surgical applications.

1.5.6 Fusion energy

Due to the ability of high-power lasers to accelerate charged particles to high energies, they are used in the development of advanced ignition schemes for fusion energy. In the block ignition process, short, high-intensity laser pulses with powers in the petawatt range induce ultrahigh acceleration of plasma blocks by the direct conversion of laser energy into macroscopic plasma motion by nonlinear (ponderomotive) forces. Such forces are used to generate high-energy electrons or ions that compress the target to a high density, which is required to achieve high gain in the generation of fusion energy. As a result, this process avoids the very complex problems of heat and instability that affect laser fusion with nanosecond pulses. A reactor based on this process would be a clean, safe, and low-cost energy source [38].

1.5.7 High-harmonic generation

Harmonic generation is a powerful technique for wavelength extension. The interaction between high-intensity ultrashort pulses and a nonlinear medium generates a burst of coherent light in the medium at odd harmonics of the pump light frequency. Such a burst of coherent light can span many octaves [39–41]. This typically occurs at optical intensities of the order of 10^{14} Wm^{-2} or higher. Details such as the harmonics generated, the harmonic power, and the cutoff wavelength depend on the nonlinear medium used, and on the characteristics of the pump light, including its wavelength, pulse duration, and pulse repetition rate. The efficiency of phase-matched harmonic generation strongly depends on the matching between the spectral content of the input fundamental light and the spectral acceptance bandwidth provided by the nonlinear medium [42]. In order to increase the conversion efficiency and to avoid harmonic pulse broadening in the regime of ultrashort lasers, the overall group velocity mismatch (GVM) between the fundamental and higher harmonic pulses in the nonlinear medium needs to be sufficiently small compared with the fundamental pulse duration, or equivalently, in the frequency domain, the spectral acceptance of the nonlinear medium needs to be sufficiently large compared with the fundamental pulse bandwidth. To meet this requirement, femtosecond pulses must be used. A particular harmonic of interest is then selected using a monochromator. These high-harmonic-generation (HHG) light sources are used in studies of the dynamics of molecular systems, time-resolved diffraction studies at the nanoscale, studies of thermal materials, and measurements of thin film properties [42–44]. A limitation of this field is average power of HHG sources, and so this is an area of active technology development.

1.6 Ultrashort-pulse-generation techniques

Ultrashort optical pulses are produced by a variety of methods. Although they differ in their technical details, each method relies on the same three key components: spectral broadening due to the nonlinear optical Kerr effect, dispersion control, and ultrabroadband amplification [10, 45–48]. In this chapter, we review state-of-the-art ultrashort pulse generation with a focus on pulse-compression schemes. In general, ultrafast light sources are either solid-state or fiber lasers. Ultrafast pulse generation is now a rapidly emerging field, because the development of optical communication system needs exceptionally high-speed transmission rates of 160 Gbps and higher. The development of ultrafast laser generation techniques in the visible and infrared regions of the electromagnetic spectrum has accelerated immense progress in numerous fields of fundamental science. In particular, high-repetition-rate ultrafast pulses have broad application prospects in laser micromachining, biomedical imaging, and photonic switching. However, it is difficult to directly generate laser pulses with so few cycles at visible and infrared wavelengths, even using the best available laser sources, because short pulses suffer from chromatic dispersion when they pass through optical elements, which changes their temporal shape. The wavelength dependence of the refractive index of the medium stretches the pulse temporally, thus lowering its peak power [49]. The evolution of ultrashort pulse-generation techniques is discussed in detail in this section.

Initially, femtosecond pulses were produced using dye lasers. In the late 1980s, the pulse duration of dye lasers was as low as 27 fs [50], which was later compressed to

6 fs [51]. At a wavelength of 600 nm, only three optical cycles fit under the FWHM of the intensity envelope of such a pulse. It took almost a decade to surpass these results with solid-state lasers. Later, solid-state quantum cascade and mode-locked lasers were utilized to generate ultrashort pulses because of their simplicity and low cost. However, as compared to mode-locked lasers, solid-state quantum cascade lasers have a number of fundamental limitations: limited average power at good beam quality and most crucially, poor energy storage capacity, which is critical for high peak power operation.

1.6.1 Mode-locking techniques

Until now, mode-locking techniques have been the technology most used for the generation of ultrashort pulses, particularly in the picosecond range [52]. In a mode-locked laser, pulse formation should start from normal noise fluctuations in the laser that initiate mode locking. In mode locking, the amplitude modulator opens and closes synchronously with the light propagating through the cavity, which causes the eigenfrequencies (or longitudinal modes) of the cavity to be phase locked when the modulation frequency is equal to the frequency spacing of the modes, generating a short pulse. In mode locking, the goal is to phase lock as many longitudinal modes as possible, because the broader the phase-locked spectrum, the shorter the pulse that can be generated. There are a number of potential operating characteristics that make mode-locking lasers particularly attractive. They have been shown to have: quantum limited noise, pulses shorter than 100 fs, more than 80 nm of bandwidth, repetition rates of up to 20 GHz, fundamental repetition rates from 100 kHz to 100 MHz, and transform-limited pulses. The first mode-locked laser was demonstrated in 1964, which suggested that as long as the unsaturated gain remains greater than the cavity losses, the modes of an incoming electromagnetic wave oscillate concurrently throughout all resonant frequencies of the cavity. In order to sustain a larger proportion of such longitudinal modes in a laser cavity, a broadband gain medium is needed. The competition within a wave packet traveling back and forth in the cavity causes the maxima to grow significantly stronger in the time domain. If the parameters are chosen correctly, a single focused pulse is created that oscillates with all of the energy of the cavity. Each time the pulse hits the output coupler mirror, a usable pulse is emitted, so that a regular pulse train leaves the laser. This is a mode-locked situation. In terms of spectral components, a short pulse is formed in the laser resonator when a fixed phase relationship is achieved between its longitudinal modes, or more precisely, between the lines in the spectrum of the laser output. The larger the number of frequency components involved, the shorter the duration of the generated pulses can be. In the steady-state femtosecond regime, dispersion and bandwidth limitations of the gain medium, mirrors, and so forth are mainly responsible for the temporal stretching of the pulse. Therefore, the pulse-shortening effect must be dominant for pulse durations ranging from nanoseconds at start-up to femtoseconds in steady-state operation. Mode-locking techniques are divided into two types based on the modulator utilized in the laser cavity, namely, active and passive mode-locking techniques (figure 1.11).

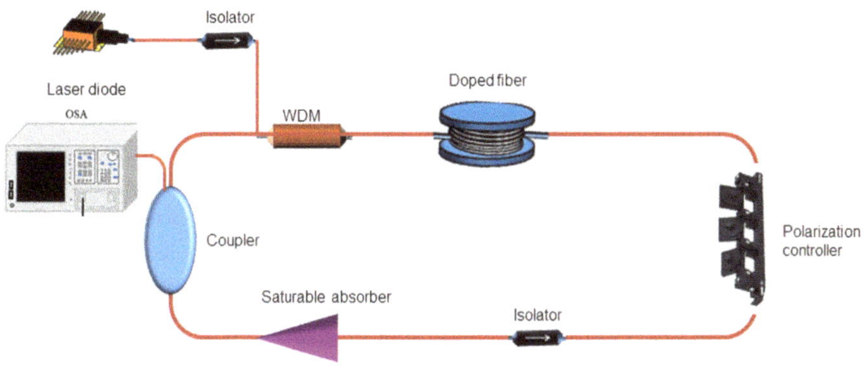

Figure 1.11. Schematic diagram of a mode-locked fiber laser system.

1.6.1.1 Active mode locking

In active mode-locking operation, a modulator is inserted into the cavity or an external optical pulse is injected to actively modulate the light wave in the fiber cavity with a period equal to the round-trip time, which controls the cavity losses or the round-trip phase change, thus amplifying a specific portion of the radiation. Active mode locking can be implemented using gain modulation (switching the pump on and off), loss modulation (the periodic decrease of loss in some element), or cavity dumping (accumulating energy in the resonator cavity and releasing it in a short burst by misaligning one mirror). This can be accomplished using either an acousto-optic or an electro-optic modulator. If the external modulation frequency is equivalent to the intermode frequency, the sidebands and longitudinal modes compete with each other in the gain medium as long as the longitudinal modes lock their phases onto the sidebands, resulting in global phase locking and amplification of the whole spectral distribution. If the modulation is synchronized with the resonator round trips, this can lead to the generation of ultrashort pulses, usually with picosecond pulse durations. This technique provides high-order harmonic mode-locking operation.

1.6.1.2 Passive mode locking

A theory of passive mode locking was developed by Haus *et al* [53]. In this model, the intracavity elements are assumed to be continuous and change on a pulse per round trip, which are treated as perturbations. Passive mode locking causes lower losses in the more intense fractions of the radiation. Lower losses are caused by fractions such as saturable absorption mode locking and Kerr lens mode locking, in which more intense radiation obtains a desirable transverse profile by self-focussing. The passive amplitude modulator is a saturable absorber which has increased transmissivity or reflectivity for high peak powers and produces self-amplitude modulation (SAM). This SAM reduces the losses for short-pulse laser operation. An optical pulse traveling through a saturable absorber in a solid-state laser is shortened by the SAM, provided that the response time of the absorber is sufficiently fast. Traditionally, dyes have been used as saturable absorbers for passive mode locking.

These dyes have been replaced by saturable absorbers obtained using Kerr or semiconductor nonlinearities. The precise control of optical nonlinearities, combined with the availability of a variety of bandgaps ranging from the visible to the infrared, makes semiconductor materials very attractive for use as saturable absorbers in solid-state lasers. Semiconductor materials typically provide an optical nonlinearity with two pronounced time constants. Intraband processes give rise to a very rapid relaxation in the 100 fs regime, while electron–hole recombination generates a slow response time in the picosecond regime. This slow response time can be reduced by several orders of magnitude using low-temperature epitaxy. For high pulse repetition rates with fundamental mode locking, very short laser resonators are required. Due to the high pulse repetition rate, the pulse energies obtained from mode-locked lasers are fairly limited—normally nanojoules or picojoules.

1.6.1.3 The influence of nonlinear effects on the mode-locking medium

Active and passive mode-locking techniques have been extensively researched in solid-state media in order to obtain high-repetition-rate pulses on the scale of a few GW. The most widely used solid-state medium is titanium-doped sapphire (Ti:Sa) crystal. The attractiveness of the Ti:Sa technology stems, in particular, from the ultrabroadband emission bandwidth of the gain material [54, 55]. With proper dispersion control, it readily enables the generation of pulses consisting of a few cycles. The scope of these techniques is limited by the stability of the pulse energy and low power scalability. This is due to the lack of available high-power pump diodes in the green wavelength region. This lack of availability of pump sources has been overcome by the development of diode lasers, which have covered an increasing number of wavelengths, allowing for a wide range of directly pumped diode lasers. This has motivated researchers to work toward the development of ultrashort pulses with millijoule pulse energies, fundamental mode beam quality, and average powers greater than one kilowatt. In order to stabilize the pulse energy, subcavities with a free spectral range have been used to match the modulation frequency. However, exact matching of the fundamental frequency, the free spectral range of the subcavity, and the modulation frequency is required. Later, additive pulse limiting (APL) and self-phase modulation (SPM) techniques were used to stabilize the pulse energy by properly adjusting the polarization bias to clamp the energy at a specific level, and by using the spectral filter inside the cavity to create more loss in the high-intensity pulse.

However, when generating femtosecond-duration pulses, the interaction of light with the cavity medium becomes nonlinear even at low pulse energies, which gives rise to phenomena such as chromatic dispersion, the Kerr effect, Raman scattering, self-phase modulation, and gain saturation. These phenomena lead to self-focussing and damage the laser medium by its interaction with the intense electric field of the light wave. The self-damage phenomena in an optical medium can be reduced by stretching out the optical pulse and thereby lowering the peak power. However, while achieving high energy density in a short pulse, the power density must not exceed the laser medium's damage threshold, which is achieved by compensating for the SPM effects. As the peak power triggers nonlinear effects, such as frequency conversion or multi-photon ionization, while the laser repetition rate determines the data acquisition rate, the combination of both is needed in various experiments, for example, in extreme ultraviolet and mid-infrared frequency comb spectroscopy [56, 57], time-resolved photo-emission electron microscopy [58], or coincidence spectroscopy [59]. Therefore, the simultaneous scaling of peak and average power is the key focus of current femtosecond technology developments.

1.6.1.4 Mode locking in fiber

The mode-locking technique has also been investigated in fiber media based on the incorporation of trivalent rare-earth ions, such as those of neodymium, erbium, and thulium, into glass hosts. The fiber itself provides the waveguide, and the availability of various fiber components minimizes the need for bulk optics and mechanical alignment. These fiber lasers can be actively or passively mode locked. In active mode locking, a modulator produces amplitude or phase modulation, while passive mode locking uses a nonlinear amplifying loop mirror, nonlinear polarization rotation, and semiconductor saturable absorbers. In passive mode locking, an intensity fluctuation acts in conjunction with fiber nonlinearity to modulate the cavity loss without external control [60]. The pulse duration of an actively mode-locked fiber laser is typically within the order of a picosecond, due to the limited response time of the modulator. In addition, the modulator can reduce the environmental stability of actively mode-locked fiber lasers. In rare-earth-doped fiber lasers, the upper-state lifetime is very long ($\approx$ ms), implying that the gain does not react significantly within the cavity round-trip time ($<0.5\ \mu$s). A fast saturable absorber is therefore required to clean up both the leading and trailing edges of the pulse. Mode locking of fiber lasers generates subpicosecond pulses. However, in fiber lasers with discrete lumped elements, a pulse undergoes appreciable nonlinear and linear evolutions. One example of this effect is dispersion management in soliton transmission. Soliton-like mode-locked pulses require a saturable absorber for stability to preferentially support pulses as opposed to continuous-wave (CW) background or dispersive waves. Furthermore, saturable absorbers play a critical role in pulse shaping. Therefore, the optimization of mode-locked laser design inevitably involves consideration of the right rate and the depth of saturation, which are important parameters of a saturable absorber. Semiconductor structures have frequently been employed for this purpose. Various optical switches, such as nonlinear loop mirrors and nonlinear polarization evolution have been more effective in mode-locking fiber lasers, because of their instantaneous responses and the fact that the short photon lifetime of fiber lasers demands a large amount of saturation.

1.7 Pulse compression

Because of the potential applications of ultrashort pulses in numerous fields of science and technology, ultrashort pulse-generation techniques have gained traction in a number of nonlinear optics domains. A well-known technique for generating ultrashort optical pulses is to employ a mode-locked fiber laser or semiconductor

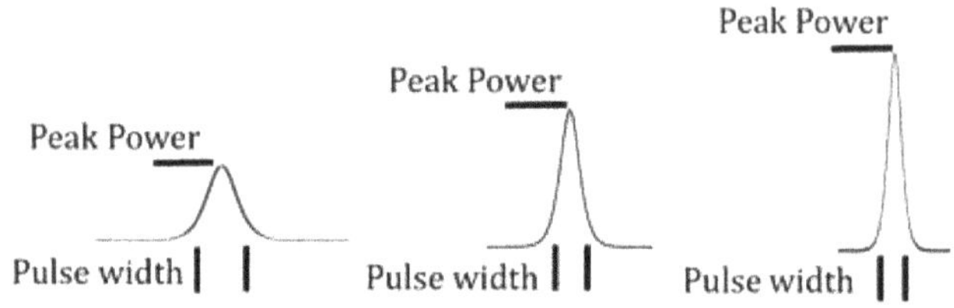

Figure 1.12. Schematic diagram of post-pulse-compression process.

laser and parametric processes. Although a mode-locked laser can generate a high-quality ultrashort optical pulse, all of the parameters involved in mode-locked operation are interdependent. For example, a small change in cavity length might cause the pulse to function with a completely new set of parameters, such as pulse width, repetition rate, carrier frequency, etc. and therefore the operation of mode-locked oscillation can be unstable. In addition, mode-locking and parametric processes necessitate more sophisticated and expensive state-of-the-art setups than those required for the generation of longer pulses. As a result, pulse-compression techniques are required to provide few-cycle pulses at any specified wavelength with greater simplicity. Similarly, many applications, such as laser fusion, need a high energy density in a short time span (figure 1.12). Pulse-compression methods are linear or nonlinear processes used to obtain bandwidth-limited pulses of extremely short duration by balancing SPM effects. Post-pulse-compression processes have been identified as the most effective method for obtaining ultrashort pulses; these work on the idea of compressing the optical energy of the electromagnetic waves in time to attain the required pulse characteristics. In order to reduce alignment and loss issues and improve the ease of coupling with transmission systems, short-pulse sources must take the form of guided-wave photonic structures. In any pulse-compression technique, the bandwidth of the coherent optical pulse restricts the degree of compression. As a result, when working with ultrashort pulses it is important to thoroughly characterize their features in order to adjust their shape to meet the application requirements.

1.7.1 Linear pulse compression

Linear pulse-compression techniques are based on the idea of chirp compensation. An instantaneous change in frequency over time of an optical pulse propagating through an optical medium is called chirp. In linear pulse-compression techniques, the temporal length of the pulse can be shortened by lowering the chirp, i.e. by flattening the spectral phase (figure 1.13).

Dechirping can be achieved by passing pulses through an optical device with an appropriate degree of chromatic dispersion (dispersion compensation), such as a grating compressor, a prism pair, an optical fiber, a chirped mirror, a chirped fiber Bragg grating, or a volume Bragg grating. The shortest possible pulse duration is then determined by the optical bandwidth of the pulse, which is unaffected by dispersive (linear) compression. In the ideal case, bandwidth-limited pulses can be obtained.

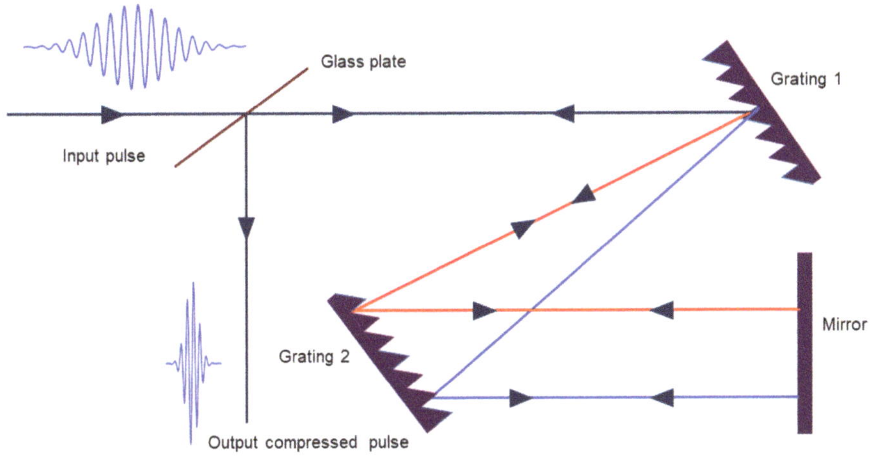

Figure 1.13. Schematic diagram of a double propagation grating compression system.

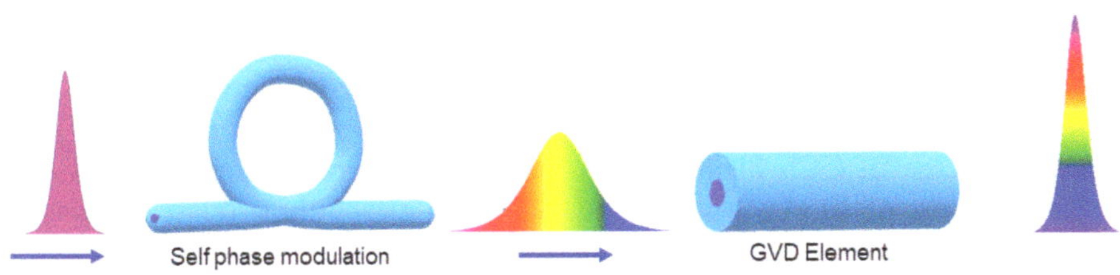

Figure 1.14. Schematic of the nonlinear pulse-compression process.

1.7.2 Nonlinear pulse compression

In a nonlinear pulse-compression system, nonlinear properties, such as self-phase modulation (SPM) (intensity-dependent phase modulation of the pulse), arise as a result of the optical Kerr effect (intensity-dependent refractive-index variation), which helps to broaden the spectrum of the pulse along with subsequent compression of the broadened spectrum. Because of the conjugate relationship between the transform-limited duration and its spectral width, a pulse becomes shortened or compressed compared to its input duration (figure 1.14). The post nonlinear compression of short pulses involves techniques such as free propagation in bulk materials, multiple-plate continuum generation, multi-pass cells, filaments, photonic crystal fibers (PCFs), hollow-core fibers, and self-compression. In most of the nonlinear pulse-compression schemes, optical fibers in the normal dispersion regime are used for spectral broadening of the pulse. For high-intensity pulses, spectral broadening has been achieved using a gas-filled hollow fiber or capillary. For soliton compression, anomalous dispersive fibers are used, but this limits the energy due to the small soliton pulse energies of typical fibers. The compressed pulse energy can be increased by amplification in a doped fiber with constant dispersion properties. Pulse compression can also occur during nonlinear frequency conversion processes, for example, those occurring in optical parametric oscillators, and in HHG. Subsequent dispersive compression can be achieved using chirped mirrors, dispersive mirrors, or bandpass filters. The selection of a pulse-compression process depends on a number of factors, including the initial and required pulse durations, the pulse energy, and the required pulse quality.

1.8 Experiments with pulse-compression techniques

This section describes the experimental realization of the various pulse-compression techniques. Initially, Nakazawa *et al* experimentally demonstrated the generation of 170–190 fs, 10 GHz pulses using a dispersion-decreasing erbium-doped fiber amplifier based on adiabatic fundamental soliton compression and adiabatic fundamental soliton-narrowing processes in a fiber with a length of 1 km [61]. Clark *et al* used a pulse-replica compression technique in a single-mode fiber with prisms and obtained 82 fs pulses with an energy of 0.4 mJ in a fiber with a length of 0.75 m [62]. With the development of PCF, Druon *et al* reported the direct compression of 110 fs pulses down to 75 fs using a PCF at a central wavelength of 1 μm with a tunability of 1–1.3 μm [63]. Ouzounov *et al* demonstrated the generation of 140 fs and 3 nJ pulses using 1.3 m microstructured large-mode-area fiber using a grating compressor [64]. Phase-compensated supercontinuum generation using a chloroform-infiltrated PCF was experimentally investigated by Le *et al.* In the case of CCl_4-filled fiber, a microfluidic pump system maintaining a constant pressure of 10 kPa was used to minimize microbubble-related effects and suppress thermal effects in the liquid under high-repetition-rate pumping conditions. CCl_4 was infiltrated into all the air holes in the cladding region of the fiber by capillary action and the pressure of the liquid pump system [65] (figure 1.15). Yang *et al* demonstrated the amplification of 120 fs pulses using an optical parametric chirped-pulse amplification system (OPCPA) consisting of two lithium triborate optical parametric preamplifiers, a potassium dihydrogen phosphate (KDP) optical parametric amplifier and neodymium-doped yttrium aluminum garnet (Nd:YAG)–Nd:glass hybrid amplifiers for the generation of 570 mJ, 155 fs pulses with a peak power of 3.67 TW [66]. Turchinovich *et al* reported a stable laser compression scheme using a 9.5m hollow-core polarization-maintaining PCF, which provided output pulses around 370 fs in duration at an energy of 4 nJ with high mode quality [67]. Pulses lasting for 77 fs with energies as large as 2.7 nJ have been generated by stretched-pulse erbium fiber lasers, although 1 nJ was produced by diode-pumped

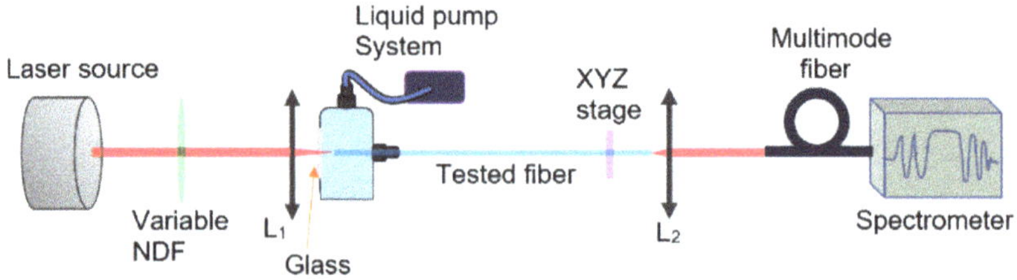

Figure 1.15. Schematic of the experimental setup used for supercontinuum generation (SCG) measurement in the CCl_4-filled fiber (Courtesy: Van Le H *et al* [65]. Reproduced with permission.).

versions [68]. Tamura *et al* experimentally demonstrated a fiber amplifier/grating pair compression system that achieved compression from 350 to 77 fs at a gain of 18 dB in an erbium-doped fiber amplifier [69]. Seidel *et al* described the compression of pulses that had an average power of 90 W and a duration of 190 fs to 70 W and 30 fs, respectively, with an increase in peak power from 18 MW to 60 MW. This compression scheme was based on the cascaded phase-mismatched quadratic nonlinearities in β-barium borate (BBO) crystals [70].

1.9 Organization of this book

Pulse compression is the process used to generate the short-duration optical pulses required for specific applications. To date, various linear and nonlinear pulse-compression processes have been proposed. In this book, an attempt is made to address the fundamental principles and limitations of the numerical modeling of the linear and nonlinear phenomena involved in the process of pulse compression in PCFs used for the generation of ultrashort pulses. The chapters are structured as follows:

Chapter 2 reviews the basic properties of optical fibers, including different types of fiber and fiber modes. It introduces PCFs, their types, and their characteristics. It explains how their optical properties are well suited for ultrafast nonlinear optics.

Chapter 3 gives a brief overview of the current numerical methods used to solve the nonlinear Schrödinger equation (NLSE) and the numerical models used to calculate PCF parameters.

Chapter 4 provides a brief description of linear and nonlinear polarization effects, followed by a short discussion of the propagation properties of light in a dielectric waveguide. It then describes the interesting coupling between linear and nonlinear effects in an optical fiber under different circumstances, and shows how they lead to the spectral broadening and soliton dynamics that facilitate the results presented in later chapters.

Chapter 5 explores existing conventional soliton-compression techniques, such as higher-order soliton compression and adiabatic pulse compression, which are used with PCFs for the generation of ultrashort pulses and discusses quality analysis of compressed pulses.

Chapter 6 discuss a pulse-compression scheme based on self-similar analysis that aims to obtain a large compression factor and minimal pedestal energy by the compression of fundamental solitons with low input pulse energies over small propagation distances. The numerical model of a PCF compressor based on self-similar properties is discussed in detail.

Chapter 7 briefly reviews the numerical simulation of simultaneous pulse compression and pedestal suppression in a tapered PCF-based nonlinear optical loop mirror (NOLM) used to model a highly efficient optical pulse compressor that generates femtosecond pulses.

Chapter 8 offers a detailed discussion of a unique cascaded soliton-compression approach using chloroform-filled cascaded PCF used for the generation of low-energy, few-cycle laser pulses. Higher-order soliton compression is introduced for the compression of femtosecond pulses.

Chapter 9 theoretically explains the generation of a tunable broadband light spectrum using a PCF by varying its temperature, which is found to have a significant impact on the process of supercontinuum generation (SCG)-induced pulse compression. Temperature-dependent fiber parameters are used to solve the NLSE, which determines the evolution of the optical pulse through the proposed fiber model.

References

[1] Maiman T 1960 Stimulated optical radiation in ruby *Nature* **187** 493–4

[2] Cerullo G, Lanzani G, Muccini M, Taliani C and De Silvestri S 1999 Real-time vibronic coupling dynamics in a prototypical conjugated oligomer *Phys. Rev. Lett.* **83** 231

[3] DeMaria A J, Stetser D A and Heynau H 1966 Self mode-locking of lasers with saturable absorbers *Appl. Phys. Lett.* **8** 174–6

[4] Krausz F and Ivanov M 2009 Attosecond physics *Rev. Mod. Phys.* **81** 163

[5] Chini M, Zhao K and Chang Z 2014 The generation, characterization and applications of broadband isolated attosecond pulses *Nat. Photon.* **8** 178–86

[6] Popmintchev T *et al* 2012 Bright coherent ultrahigh harmonics in the KeV x-ray regime from mid-infrared femtosecond lasers *Science* **336** 1287–91

[7] Kraus P M, Zürch M, Cushing S K, Neumark D M and Leone S R 2018 The ultrafast x-ray spectroscopic revolution in chemical dynamics *Nat. Rev. Chem* **2** 82–94

[8] Lépine F, Ivanov M Y and Vrakking M J J 2014 Attosecond molecular dynamics: fact or fiction? *Nat. Photon.* **8** 195–204

[9] Fattahi H *et al* 2014 Third-generation femtosecond technology *Optica* **1** 45–63

[10] Shirakawa A, Sakane I, Takasaka M and Kobayashi T 1999 Sub-5-fs visible pulse generation by pulse-front-matched noncollinear optical parametric amplification *Appl. Phys. Lett.* **74** 2268–70

[11] Yefet S and Peer A 2013 A review of cavity design for Kerr lens mode-locked solid-state lasers *Appl. Sci.* **3** 694–724

[12] Falconieri M, Marrocco M, Merla C, Gagliardi S, Rondino F and Ghezelbash M 2019 Characterization of supercontinuum generation in a photonic crystal fiber for uses in multiplex cars microspectroscopy *J. Raman Spectrosc.* **50** 1287–95

[13] Crooker S A 2002 Fiber-coupled antennas for ultrafast coherent terahertz spectroscopy in low temperatures and high magnetic fields *Rev. Sci. Instrum.* **73** 3258–64

[14] Babushkin I, Tajalli A, Sayinc H, Morgner U, Steinmeyer G and Demircan A 2017 Simple route toward efficient frequency conversion for generation of fully coherent supercontinua in the mid-IR and UV range *Light: Sci. Appl.* **6** e16218

[15] Woodward J T, Smith A W, Jenkins C A, Lin C, Brown S W and Lykke K R 2009 Supercontinuum sources for metrology *Metrologia* **46** S277

[16] Pouli P, Bounos G, Georgiou S and Fotakis C 2007 Femtosecond laser cleaning of painted artefacts; is this the way forward? *Lasers in the Conservation of Artworks* (Berlin: Springer) pp 287–93

[17] Humbert G *et al* 2006 Supercontinuum generation system for optical coherence tomography based on tapered photonic crystal fibre *Opt. Express* **14** 1596–603

[18] Israelsen N M, Petersen C R, Barh A, Jain D, Jensen M, Hannesschläger G, Tidemand-Lichtenberg P, Pedersen C, Podoleanu A and Bang O 2019 Real-time high-resolution mid-infrared optical coherence tomography *Light: Sci. Appl.* **8** 1–13

[19] Tearney G J, Brezinski M E, Bouma B E, Boppart S A, Pitris C, Southern J F and Fujimoto J G 1997 In vivo endoscopic optical biopsy with optical coherence tomography *Science* **276** 2037–9
[20] Chichkov B N, Momma C, Nolte S, Von Alvensleben F and Tünnermann A 1996 Femtosecond, picosecond and nanosecond laser ablation of solids *Appl. Phys.* A **63** 109–15
[21] Krüger J, Kautek W, Lenzner M, Sartania S, Spielmann C and Krausz F 1998 Laser micromachining of barium aluminium borosilicate glass with pulse durations between 20 fs and 3 ps *Appl. Surf. Sci.* **127** 892–8
[22] Banerjee S *et al* 2015 Compact source of narrowband and tunable x-rays for radiography *Nucl. Instrum. Methods Phys. Res.* B **350** 106–11
[23] Chen L R, Benjamin S D, Smith P W E and Sipe J E 1998 Applications of ultrashort pulse propagation in Bragg gratings for wavelength-division multiplexing and code-division multiple access *IEEE J. Quantum Electron.* **34** 2117–29
[24] De Souza E A, Nuss M C, Knox W H and Miller D A B 1995 Wavelength-division multiplexing with femtosecond pulses *Opt. Lett.* **20** 1166–8
[25] Steen W M and Mazumder J 2010 *Laser Material Processing* (Berlin: Springer)
[26] Liu X, Du D and Mourou G 1997 Laser ablation and micromachining with ultrashort laser pulses *IEEE J. Quantum Electron.* **33** 1706–16
[27] Vorobyev A Y and Guo C 2005 Enhanced absorptance of gold following multipulse femtosecond laser ablation *Phys. Rev.* B **72** 195422
[28] Korte F, Nolte S, Chichkov B N, Bauer T, Kamlage G, Wagner T, Fallnich C and Welling H 1999 Far-field and near-field material processing with femtosecond laser pulses *Appl. Phys.* A **69** S7–11
[29] Jiang L, Fang J, Cao Q, Zhang K, Wang P, Yu Y, Huang Q and Lu Y 2014 Femtosecond laser high-efficiency drilling of high-aspect-ratio microholes based on free-electron-density adjustments *Appl. Opt.* **53** 7290–5
[30] Groenendijk M 2008 Fabrication of super hydrophobic surfaces by fs laser pulses: how to produce self-cleaning surfaces *Laser Tech.* J **5** 44–7
[31] Tokarev V N, Cheshev E A, Bezotosnyi V V, Khomich V Y, Mikolutskiy S I and Vasil'yeva N V 2015 Optimization of plasma effect in laser drilling of high aspect ratio microvias *Laser Phys.* **25** 056003
[32] Sakimoto T, Rosenblatt M I and Azar D T 2006 Laser eye surgery for refractive errors *Lancet* **367** 1432–47
[33] Toth R, Kieffer J C, Fourmaux S, Ozaki T and Krol A 2005 In-line phase-contrast imaging with a laser-based hard x-ray source *Rev. Sci. Instrum.* **76** 083701
[34] Wenz J, Schleede S, Khrennikov K, Bech M, Thibault P, Heigoldt M, Pfeiffer F and Karsch S 2015 Quantitative x-ray phase-contrast microtomography from a compact laser-driven betatron source *Nat. Commun.* **6** 1–6
[35] Yogo A *et al* 2009 Application of laser-accelerated protons to the demonstration of DNA double-strand breaks in human cancer cells *Appl. Phys. Lett.* **94** 181502
[36] Linz U and Alonso J 2007 What will it take for laser driven proton accelerators to be applied to tumor therapy? *Phys. Rev. Spec. Top. Accel. Beams* **10** 094801
[37] Cole J M *et al* 2015 Laser-wakefield accelerators as hard x-ray sources for 3d medical imaging of human bone *Sci. Rep.* **5** 1–7
[38] Hora H 2015 Clean boron fusion using extreme laser pulses *SPIE News Letter*

[39] Zhu H, Wang T, Zheng W, Yuan P, Qian L and Fan D 2004 Efficient second harmonic generation of femtosecond laser at 1 μm *Opt. Express* **12** 2150–5
[40] Spielmann Ch Burnett N H, Sartania S, Koppitsch R, Schnürer M, Kan Ch, Lenzner M, Wobrauschek P and Krausz F 1997 Generation of coherent x-rays in the water window using 5-femtosecond laser pulses *Science* **278** 661–4
[41] Ghimire S, DiChiara A D, Sistrunk E, Agostini P, DiMauro L F and Reis D A 2011 Observation of high-order harmonic generation in a bulk crystal *Nat. Phys.* **7** 138–41
[42] Baltuška A, Fuji T and Kobayashi T 2002 Controlling the carrier-envelope phase of ultrashort light pulses with optical parametric amplifiers *Phys. Rev. Lett.* **88** 133901
[43] Li J, Lu J, Chew A, Han S, Li J, Wu Y, Wang H, Ghimire S and Chang Z 2020 Attosecond science based on high harmonic generation from gases and solids *Nat. Commun.* **11** 1–13
[44] Chang Z, Rundquist A, Wang H, Murnane M M and Kapteyn H C 1997 Generation of coherent soft x rays at 2.7 nm using high harmonics *Phys. Rev. Lett.* **79** 2967
[45] Morgner U, Kärtner F X, Cho S-H, Chen Y, Haus H A, Fujimoto J G, Ippen E P, Scheuer V, Angelow G and Tschudi T 1999 Sub-two-cycle pulses from a kerr-lens mode-locked Ti: sapphire laser *Opt. Lett.* **24** 411–3
[46] Sutter D H, Steinmeyer G, Gallmann L, Matuschek N, Morier-Genoud F, Keller U, Scheuer V, Angelow G and Tschudi T 1999 Semiconductor saturable-absorber mirror-assisted Kerr-lens mode-locked Ti: sapphire laser producing pulses in the two-cycle regime *Opt. Lett.* **24** 631–3
[47] Baltuška A, Wei Z, Pshenichnikov M S, Wiersma D A and Szipöcs R 1997 All-solid-state cavity-dumped sub-5-fs laser *Appl. Phys.* B **65** 175–88
[48] Nisoli M, Stagira S, De Silvestri S, Svelto O, Sartania S, Cheng Z, Lenzner M, Spielmann Ch and Krausz F 1997 A novel-high energy pulse compression system: generation of multi-gigawatt sub-5-fs pulses *Appl. Phys. B* **65** 189–196
[49] Le T, Hofer M, Cheng Z, Stingl A, Darmo J, Kelly D P and Unterrainer K 2008 Advances in fiber delivery of ultrashort pulses at 800 nm *Solid State Lasers XVII: Technology and Devices* (Bellingham, WA: International Society for Optics and Photonics)
[50] Valdmanis J and Fork R 1986 Design considerations for a femtosecond pulse laser balancing self phase modulation, group velocity dispersion, saturable absorption, and saturable gain *IEEE J. Quantum Electron.* **22** 112–8
[51] Fork R L, Brito Cruz C H, Becker P C and Shank C V 1987 Compression of optical pulses to six femtoseconds by using cubic phase compensation *Opt. Lett.* **12** 483–5
[52] Han Y, Guo Y, Gao B, Ma C, Zhang R and Zhang H 2020 Generation, optimization, and application of ultrashort femtosecond pulse in mode-locked fiber lasers *Prog. Quantum Electron.* **71** 100264
[53] Haus H A, Fujimoto J G and Ippen E P 1992 Analytic theory of additive pulse and Kerr lens mode locking *IEEE J. Quantum Electron.* **28** 2086–96
[54] Krausz F, Fermann M E, Brabec T, Curley P F, Hofer M, Ober M H, Spielmann C, Wintner E and Schmidt A J 1992 Femtosecond solid-state lasers *IEEE J. Quantum Electron.* **28** 2097–122
[55] French P M W 1995 The generation of ultrashort laser pulses *Rep. Prog. Phys.* **58** 169
[56] Gohle C, Udem T, Herrmann M, Rauschenberger J, Holzwarth R, Schuessler H A, Krausz F and Hänsch T W 2005 A frequency comb in the extreme ultraviolet *Nature* **436** 234–7
[57] Schliesser A, Picqué N and Hänsch T W 2012 Mid-infrared frequency combs *Nat. Photon.* **6** 440–9

[58] Mikkelsen A *et al* 2009 Photoemission electron microscopy using extreme ultraviolet attosecond pulse trains *Rev. Sci. Instrum.* **80** 123703
[59] Liu Y, Tschuch S, Dürr M, Rudenko A, Moshammer R, Ullrich J, Siegel M and Morgner U 2007 Towards non-sequential double ionization of Ne and Ar using a femtosecond laser oscillator *Opt. Express* **15** 18103–10
[60] Hofer M, Ober M H, Haberl F and Fermann M E 1992 Characterization of ultrashort pulse formation in passively mode-locked fiber lasers *IEEE J. Quantum Electron.* **28** 720–8
[61] Nakazawa M, Yoshida E, Kubota H and Kimura Y 1994 Generation of a 170 fs, 10 GHz transform-limited pulse train at 1.55 μm using a dispersion-decreasing, erbium-doped active soliton compressor *Electron. Lett.* **30** 2038–40
[62] Clark S W, Ilday F and Wise F W 2001 Fiber delivery of femtosecond pulses from a Ti: sapphire laser *Opt. Lett.* **26** 1320–2
[63] Druon F, Sanner N, Lucas-Leclin G, Georges P, Hansen K P and Petersson A 2003 Self-compression and Raman soliton generation in a photonic crystal fiber of 100-fs pulses produced by a diode-pumped Yb-doped oscillator *Appl. Opt.* **42** 6768–70
[64] Ouzounov D G, Moll K D, Foster M A, Zipfel W R, Webb W W and Gaeta A L 2002 Delivery of nanojoule femtosecond pulses through large-core microstructured fibers *Opt. Lett.* **27** 1513–5
[65] Van Le H *et al* 2021 Low pump power coherent supercontinuum generation in heavy metal oxide solid-core photonic crystal fibers infiltrated with carbon tetrachloride covering 930-2500 nm *Opt. Express* **29** 39586–600
[66] Yang X, Xu Z-z, Leng Y-x, Lu H-h, Lin L-h, Zhang Z-q, Li R-x, Zhang W-q, Yin D-j and Tang B 2002 Multiterawatt laser system based on optical parametric chirped pulse amplification *Opt. Lett.* **27** 1135–7
[67] Turchinovich D, Liu X and Lægsgaard J 2008 *Opt. Express* **16** 14004–14
[68] Tamura K, Ippen E P, Haus H A and Nelson L E 1993 77 -fs pulse generation from a stretched-pulse mode-locked all-fiber ring laser *Opt. Lett.* **18** 1080–2
[69] Tamura K and Nakazawa M 1996 Pulse compression by nonlinear pulse evolution with reduced optical wave breaking in erbium-doped fiber amplifiers *Opt. Lett.* **21** 68–70
[70] Seidel M, Brons J, Arisholm G, Fritsch K, Pervak V and Pronin O 2017 Efficient high-power ultrashort pulse compression in self-defocusing bulk media *Sci. Rep.* **7** 1–8
[71] Turgut B and Demir T 2016 The new landmarks, findings and signs in optical coherence tomography *New Front. Ophth.* **2** 131–6
[72] Ancona A, Röser F, Rademaker K, Limpert J, Nolte S and Tünnermann A 2008 High speed laser drilling of metals using a high repetition rate, high average power ultrafast fiber CPA system *Opt. Expr.* **16** 8958–68
[73] Tan D T H, Agarwal A M and Kimerling L C 2015 Nonlinear photonic waveguides for on-chip optical pulse compression *Laser Photon. Rev.* **9** 294–308

Chapter 2

Photonic crystal fiber

This chapter provides an introduction to optical fiber, specifically, photonic crystal fiber. It discusses its types and the various parameters involved in the guidance of light through the fiber as well as the optical properties of PCF and how they are well suited for ultrafast nonlinear optics.

2.1 Optical fiber

Fibers are dielectric waveguides that guide light from one end to the other. An optical fiber is considered to be one of the most significant tools in the generation of ultrashort pulses because of its low transmission loss and high optical damage threshold. In optical signal transmission, fibers are used rather than metal wires because they are immune to electromagnetic interference, a problem from which metal wires suffer excessively. Because of this property, optical fibers have found a wide range of uses in fiber-optic communications, where they permit transmission over long distances and at higher bandwidths (data rates) than electrical cables. Fiber-optic systems are widely employed in broadband communications systems, optical imaging, sensors, materials processing, etc. In generating and transmitting different types of optical pulses, it is important to understand the physical properties of optical fibers, such as their waveguide characteristics, chromatic dispersion, loss, and nonlinearity in order to take full advantage of their potential.

2.2 Guiding mechanism of optical fiber

In 1870, John Tyndall provided experimental proof for the bending of light by directing light through a falling stream of water, as shown in figure 2.1. This proves that light need not travel in a straight line; it can be confined to a curved stream of water. This kind of phenomenon is known as total internal reflection (TIR). Based on Tyndall's study of how light can be confined to a curved stream of water, the propagation of light inside optical fiber has been explained using the phenomenon of TIR, which is even more efficient than mirrored reflection; it reflects more than

doi:10.1088/978-0-7503-2686-5ch2

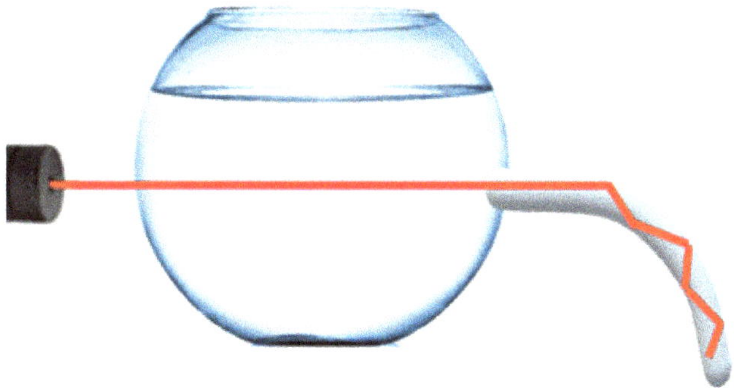

Figure 2.1. Concept of light guidance through a water tank demonstrated by Tyndall's experiment.

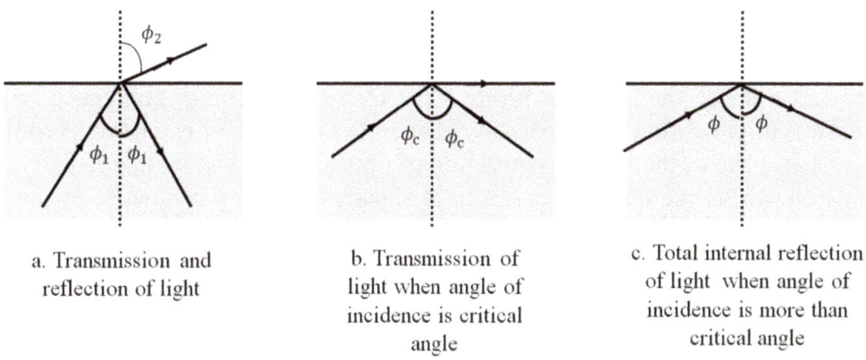

Figure 2.2. Representations of the phenomena of (a) refraction, (b) reflection, (c) TIR.

99.9% of the light. In transparent materials, TIR depends on the property called the refractive index (RI), which is determined by the ratio of the speed of light in vacuum to the speed of light in the medium. This refractive index value controls the speed of light in the propagating medium. As a result, when a light ray is incident on the interface between a high-refractive-index dielectric medium (n_1) and a lower-index dielectric medium (n_2) at an angle of ϕ_1, refraction occurs at an angle of ϕ_2, i.e. there is a change in the direction of light propagation, as illustrated in figure 2.2(a). In addition, a small amount of light is reflected back into the propagating medium, as shown in figure 2.2(a). The relationship between the angle of incidence ϕ_1, the angle of refraction ϕ_2, and the refractive indices of the dielectric media is given by Snell's law of refraction, as follows:

$$n_1 \sin(\phi_1) = n_2 \sin(\phi_2). \tag{2.1}$$

Here, a value of n_1 that is greater than n_2 leads to larger angle of refraction (ϕ_2) than the angle of incidence (ϕ_1). At $\phi_2 = 90°$, the refracted ray follows the line of the interface between the two dielectrics. The angle of incidence at this time is known as the critical angle (ϕ_c), which is given by

$$\sin(\phi_c) = \frac{n_2}{n_1}. \tag{2.2}$$

As the angle of incidence exceeds the critical angle, the light incident on the interface is reflected back into the original medium as shown in figure 2.2(b). For further increments in the incident angle, total reflection of light occurs at the interface (it does not enter the second medium) with a reflection angle equal to the incident angle, as shown in figure 2.2(c). This event is termed TIR.

2.3 Optical fiber construction

Optical fiber has a very simple structure which consists of two parts: an inner layer of high-refractive-index dielectric glass as a core and an outer layer of lower-refractive-index glass as a cladding. Because of the RI difference between the core (n_1) and the cladding (n_2), light guidance via TIR is possible when light is allowed to enter the fiber at an angle of incidence within the range of acceptance angles. Due to this guiding property, optical fiber is also called an 'optical waveguide.' In addition to these layers, secondary cladding is also present, which does not affect light propagation inside the fiber but protects the fiber from damage and shock, giving it mechanical protection and flexibility. Hence, this outer layer is termed as coating. Figure 2.3(a) shows the basic construction of optical fiber. The middle region of the fiber is known as the core; it carries light signals over long distances. In most of optical fibers the range of the core diameter varies from 8 ~ 62.5 μm. The second layer of the optical fiber is the cladding, and its diameter usually lies in the range of 125 μm. To put this size into perspective, a human hair is around 75 μm. Similarly, the buffer coating is made up of acrylic or nylon and its diameter varies from 250 μm to 900 μm. At the boundary between these two media, light is reflected by TIR and trapped in the core while it is traveling along the fiber. Figure 2.4 illustrates light guidance inside an optical fiber based on the phenomenon of TIR at the core–cladding interface. When light is allowed to enter into the fiber at an angle of incidence (i), and ϕ from an imaginary line perpendicular to the surface passing from the core to the cladding, it bends away from the normal line. At one particular angle, larger than the critical angle $\phi_c\left[= \sin^{-1}\left(\frac{n_2}{n_1}\right)\right]$, at the core–cladding interface, the

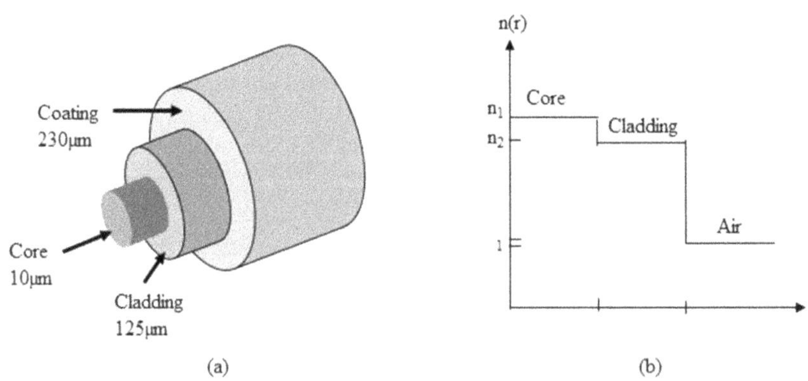

Figure 2.3. (a) Cross-sectional representation of an optical fiber, (b) variation in the refractive index profile of optical fiber.

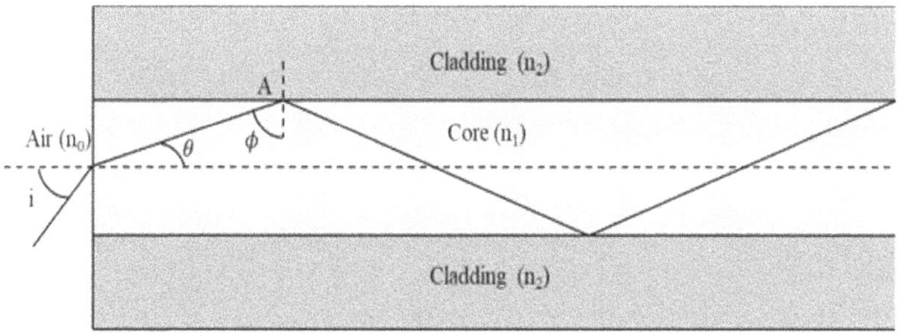

Figure 2.4. Schematic representation of the incidence of a light ray on an interface at an angle larger than the critical angle and trapping of light inside the fiber due to TIR.

refracted light does not pass into the cladding, but instead experiences TIR and travels inside the core of the fiber.

2.4 Modes in optical fiber

In optical fiber, light waves are guided in the form of distinct patterns, which are known as *modes*. In short, a mode is the number of paths of in-phase electromagnetic (EM) waves in an optical fiber. The mathematical modeling of mode theory in optical fiber is derived from Maxwell's equations. Since light is an EM wave, waves incident on the fiber at an angle greater than the critical angle are transmitted through it via internal reflections. The light rays that propagate inside the fiber are categorized as: (i) axial rays, or (ii) zig-zag rays, which are totally internally reflected at the walls of the fiber. This leads to the occurrence of a phase shift between the propagating waves. Thus, the waves that are in phase along a certain zig-zag path interfere constructively and become intensified, while the waves in the other paths are out of phase and fade due to destructive interference. The distinct pattern of light inside the fiber depends on the diameter of the core, the transmitted light wavelength and the refractive index change. Generally, modes are represented by m, which is known as the order number (figure 2.5 and figure 2.6).

A light ray propagating along the fiber axis is called a zeroth-order ray or the fundamental mode. There may be N possible modes of propagation inside the fiber. Among these, the fiber only selects certain modes of light based on the phase conditions; the selection also depends on fiber parameters such as the numerical aperture (NA), the fiber's core diameter and the operating wavelength.

2.5 Normalized frequency (V number) of a core

Another parameter that helps to characterize a fiber is the V number or normalized frequency of the fiber. The normalized frequency is a dimensionless parameter which relates three important variables used in the design of a fiber: namely, the core radius (a), the relative refractive index difference (Δ), and the operating wavelength (λ):

$$V = \frac{2\pi a}{\lambda}\sqrt{n_1^2 - n_2^2}. \tag{2.3}$$

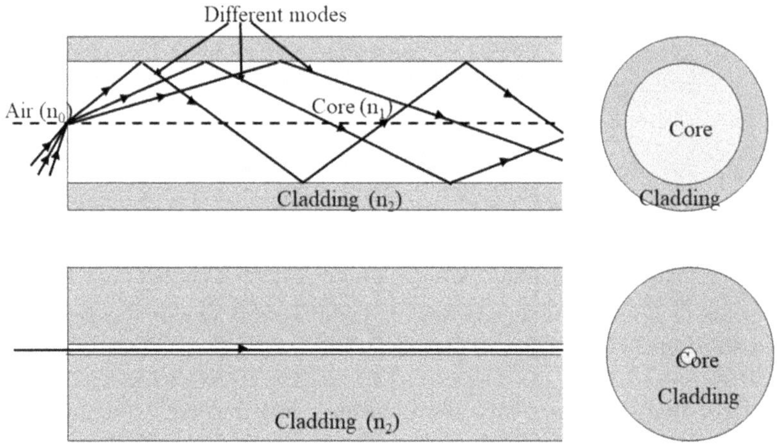

Figure 2.5. The light ray paths in (a) a large-diameter fiber and (b) a small-diameter fiber.

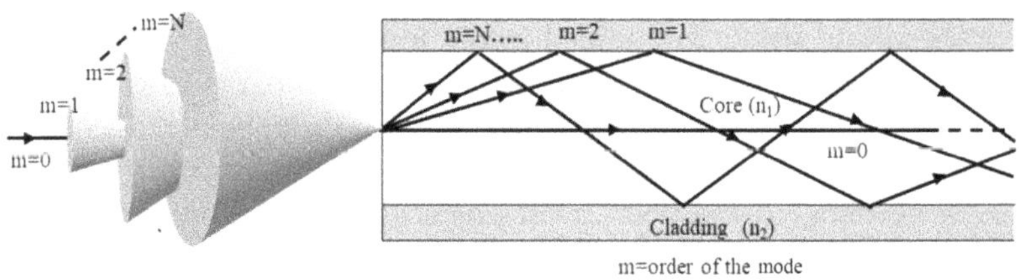

Figure 2.6. Modal propagation of light.

The NA is given by

$$NA = \sqrt{n_1^2 - n_2^2} = n_1\sqrt{2\Delta} \tag{2.4}$$

$$V = \frac{2\pi a}{\lambda}(NA) = \frac{2\pi a}{\lambda} n_1\sqrt{2\Delta}. \tag{2.5}$$

The maximum number of modes N_m may be obtained using the relation:

$$N_m \approx \frac{V^2}{2} \tag{2.6}$$

$$N_m \approx \frac{\left(\text{Diameter of the core} \times NA \times \dfrac{\pi}{\lambda}\right)^2}{2}. \tag{2.7}$$

Modes are also characterized by the V number. For $V < 2.405$, fiber allows only one mode to propagate through it; such fiber is called single-mode fiber (SMF). However, for $V > 2.405$, fiber can support any number of modes traveling inside it, each of which propagates at its own velocity; such fiber is called multimode fiber

2.6 Transmission window

In general, standard optical fibers are made of fused silica material which transmits light in the wavelength range of 0.2 μm to 2.3 μm. Initially, the wavelength band around 800–900 nm (the first communication window or short-wavelength band) was used for optical fiber communications in the early 1980s because of the dip in fiber absorption at that wavelength at the time. Later, with the development of doped silica optical fiber manufacturing, a remarkable increase in transmission characteristics was achieved, notably a decrease in pulse distortion and attenuation, while optical signals traveling through fiber transformed the field of fiber optics. Optical pulse around an operating wavelength of 1300 nm pass through SMF with no pulse broadening. Hence, most fiber-optic communications systems operate at a transmission window of around 1300 nm (the second window or medium-wavelength band). Later, the lowest loss of silica achieved using the 1550 nm wavelength band led to the modeling of dispersion-shifted fibers with negligible dispersion around the 1550 nm band (the third window or long-wavelength band), giving us fibers with the lowest loss of about 0.26 dB m^{-1} and nearly zero dispersion (figure 2.7). Beyond that, nonlinear effects are used to counteract the effects of dispersion in fiber. Because of the substantial nonlinear effects caused by the small cross-sectional areas of the beams guided by fiber, even low powers can result in high-intensity light beams in optical fiber, which makes short-pulse fiber lasers more advantageous than bulky solid-state lasers, especially when their compact size and freedom from misalignment are considered.

2.7 Pulse compression in optical fiber

Ultrashort pulses are usually generated by the compression of mode-locked laser output in optical fiber. The propagation of pulses through a fiber is usually described

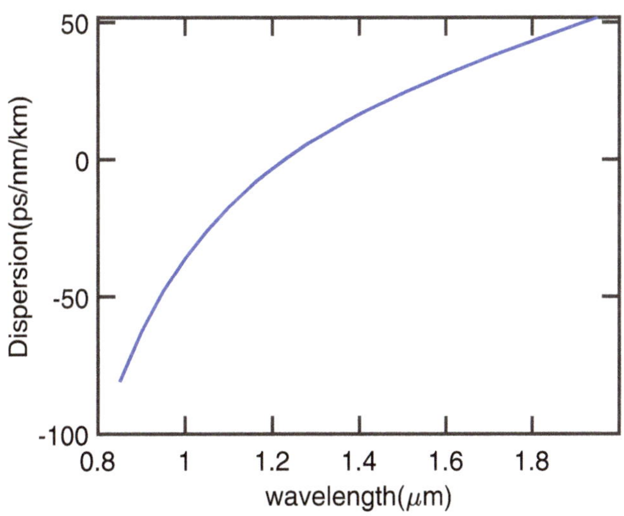

Figure 2.7. Dispersion of SMF.

by two parameters, namely the nonlinear length L_{NL} and the dispersion length L_D. If L_D is much smaller than L_{NL}, the pulses are linearly stretched or compressed, depending on their initial chirp. The fiber behaves as more or less a transparent bulk. In contrast, if L_{NL} is shorter than L_D, the pulses suffer spectral broadening or narrowing and lose their capability to exit the fiber with the initial pulse duration. In order to avoid nonlinearities or to keep them as small as possible, one has to limit the peak power P to make L_{NL} larger than L_D. This ensures that linear stretching happens in a much shorter timescale than the timescale in which nonlinear interactions can essentially evolve.

The delivery of ultrashort pulses using optical fiber opens up remarkable opportunities to simplify optical setups and reach inaccessible regions. Compared to solid-state pulse-generation techniques, fiber schemes are much more compact, cheap, and easy to handle. Moreover, light distribution, e.g. for optical communications, is commonly operated using fiber optics in which laser pulses coming directly from the fiber are more advantageous in terms of coupling quality. For pulses with relatively low intensities, silica fibers can be used for compression. However, laser pulses traveling through optical fibers are affected by material dispersion and nonlinear effects due to their confinement to rather small core of the waveguide. For example, the material dispersion of bulk silica generates considerable temporal pulse broadening, which limits the possible bandwidth of this technique.

A much better approach for enhancing nonlinear effects is to increase the effective core diameter, and hence A_{eff}, which allows more powerful pulse propagation. The mode area of the fiber can be customized to allow the propagation of only a single mode, which can help to clean the spatial beam profile at the output. Furthermore, in order to balance the impact of dispersion, soliton generation in optical fiber has been introduced. However, the realization of soliton generation in optical fiber needs hundreds of kilometers long optical fiber to counterbalance the effect of dispersion due to SPM. The presence of nonlinear effects such as SPM in conventional solid-core optical fiber has been used for spectral broadening of the pulse. For example, a short length of SMF combined with a chirped mirror compressor enabled a reduction of pulse duration to 4 fs. As a result, generating solitons over relatively small distances in conventional solid-core optical fiber is practically impossible. However, scientists have recently been able to generate solitons in photonic crystal fiber (PCF) with lengths of the order of centimeters. This is possible because the second-order dispersion (SOD) coefficient in PCF is hundreds of orders of magnitude greater than that of standard telecommunications fiber.

2.8 Photonic crystal fiber

A new kind of quasi one-dimensional optical fiber based on the properties of photonic crystal has revolutionized the field of fiber optics by offering a new light-guidance technique that was not available using ordinary optical fibers [1, 2]. The invention of PCF has enabled disruptive technological advances to take place. Not only does PCF offer improved optical properties over conventional step-index fibers, it uniquely allows strong guidance within the core. A photonic crystal is a microstructured material in which low- and high-index materials are regularly patterned in one, two, or three dimensions with a characteristic length or period of the order of the wavelength. Several names have been coined to refer these fibers, namely holey fiber, microstructured optical fiber, and photonic bandgap fiber (PBGF). While the names 'PCF,' 'holey fiber,' and 'microstructured optical fiber' originate from a structural point of view, 'PBGF' is based on PCF's optical property. PCFs produce their light-guidance effect through a patterning of tiny holes, which run along the entire length of the fiber in the cladding region. These help to achieve large-refractive-index modulation by tailoring the fiber's waveguide dispersion [3]. In the simplest PCFs, a core of fused solid silica is surrounded by many transversely microstructured air channels or voids; their microstructural scale is comparable to the wavelength of the electromagnetic radiation guided by the fiber, thus lowering its effective refractive index and allowing guidance by TIR. PCFs can therefore be made using a single material, i.e. without chemical doping. These fibers exhibit translational symmetry along the longitudinal direction of the fibers (i.e. along the z-axis). Thus, they have a periodic variation of their refractive indices in the plane perpendicular to the direction of light propagation. A typical PCF with circular air holes in a hexagonal arrangement is shown in figure 2.8. The characteristics of PCFs can be adjusted by changing the parameters of the holes. The flexible structure of PCF gives it the advantages of high nonlinearity, endless single-mode transmission, controllable mode field area and dispersion, high birefringence, etc [4–10].

In PCFs, light is guided in the core region by the index difference or photonic bandgap effect, which is a result of the structural modifications in the fiber [1, 2]. The key structural parameters of this fiber are the hole diameter (d), the pitch (center-to-center hole spacing) (Λ), the core diameter, and the number of air-hole rings. Changes in the structural parameters can easily be tailored during the fiber drawing process, providing better design flexibility and unique physical properties compared to conventional optical fibers. In PCF the difference in refractive index between the core and the cladding is significantly greater than in standard silica fiber, and the difference in refractive index may be increased or decreased by adjusting the size of the air holes. Another important benefit of PCFs is that they have smaller intrinsic

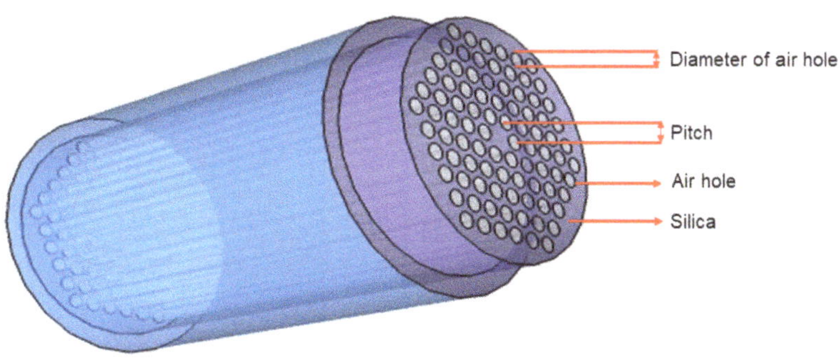

Figure 2.8. Schematic diagram of a PCF; the air-hole diameter is d and the distance between air holes is Λ.

losses than regular silica fiber. Other than the periodic air-hole design in PCF's, filling either the core or the cladding with liquid to attain an appropriate refractive index has also been extensively investigated for light propagation. This significant variation has enabled the use of PCFs in applications such as wavelength conversion using four-wave mixing (FWM), supercontinuum generation [11], optimization of pump spectra to achieve flat Raman gain, minimization of a PCF amplifier's noise figure, Raman lasing characteristics, narrow or broad bandpass filters, excellent carriers for polarization beam splitters [12], and so on.

2.8.1 Types of photonic crystal fiber

PCFs can be classified into two categories based on their guiding principle, namely PBGF and index-guiding PCF.

(i) Photonic bandgap fiber

A PBGF is an optical waveguide in which microstructured cladding provides a one or two-dimensional photonic bandgap that confines light to the fiber core [13, 14]. If the frequency of the light is within the bandgap of the two-dimensional photonic crystal formed by the periodic cladding, light guidance is attained by coherent Bragg scattering, in which light at wavelengths within well-defined stop bands is prohibited from propagating in the photonic crystal's cladding and is confined to the core. A hollow-core PCF consists of a micrometer-scale air hole surrounded by a photonic crystal cladding formed by hundreds of capillaries running along the length of the fiber; its air-filling factor is 90%. In PBGFs, the photonic crystal cladding acts as a mirror and more than 99% of the optical power is located in the air and not in the glass. The hollow core of PBGF acts as a photonic barrier so that light with wavelengths corresponding to the bandgaps cannot escape the core and is thus guided along the fiber with low loss [15]. PBGF exhibits minimal nonlinearity and negligible dispersion. About a decade ago, advancements in fabrication technologies led to realization of PBGFs and the study of their properties. Figure 2.9(a) shows the PBGF designed by a group at the University of Bath. Since this type of fiber exhibits

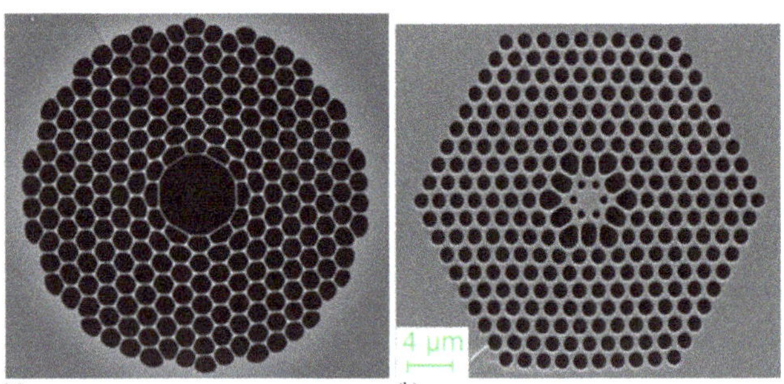

Figure 2.9. (a) SEM image of a PBGF in which light is guided by the photonic bandgap effect (courtesy: Poletti *et al* [51] reproduced with permission). (b) SEM image of a PCF in which the light is guided by TIR (courtesy: Le H V *et al* [52] reproduced with permission).

minimal nonlinearity and negligible dispersion, there are numerous applications of PBGFs namely sensors, marking, gas spectroscopy, low-distortion data transmission, laser surgery, and compression of femtosecond fiber laser pulses [16–20].

(ii) Index-guiding PCF

Light guidance in index-guiding PCF is similar to that in conventional fibers, i.e. through modified TIR, which occurs at the interface between a high-index core and a low-index cladding. As shown in figure 2.9(b) and figure 2.10, index-guiding PCF consists of a solid core enclosed in a two-dimensional photonic crystal that contains an array of air holes spaced at intervals of a few micrometers and organized in a hexagonal pattern. This solid core is formed by introducing a defect (the omission of a single air hole) into a 2D photonic crystal. In a conventional fiber, the core is constructed of one kind of material and the cladding is made of another material with a slightly lower refractive index in order to produce the needed step-index profile. By contrast, index-guiding PCF is made up of a single material and depends on subtle variations in refractive index created by the air-hole diameter (d) and the distance between air holes, known as the pitch (Λ); it does not have an explicit boundary between the core and cladding areas. The guiding property of index--guiding PCF is qualitatively controlled by the air-filling fraction or the ratio d/Λ. The index difference between the core and the cladding, as well as the effective core area, may readily be modified to achieve the desired attributes for the intended applications. Furthermore, by varying the geometrical dimensions, the zero-dispersion wavelength (ZDW) of an index-guiding PCF may be tuned to any regime, which helps to attain infinite single-mode propagation in a silica PCF, even in the visible wavelength zone. It also has high nonlinearity, high dispersion, and low loss. Since nonlinearity and chromatic dispersion are impacted not only by material properties but also by fiber design, index-guiding PCFs have unquestionably caused a major revolution in the area of nonlinear optics [21–23].

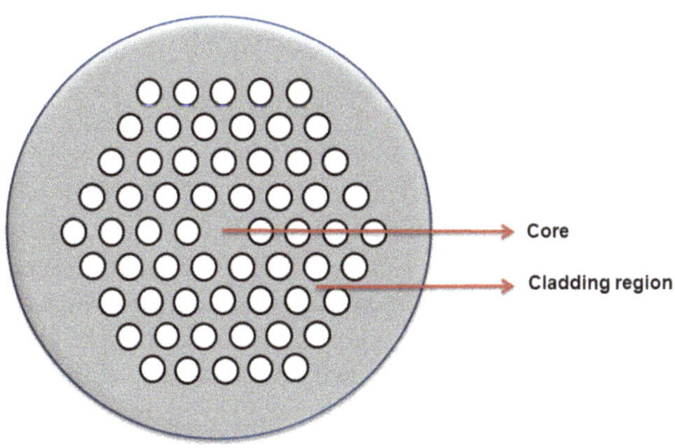

Figure 2.10. Cross-sectional view of an index-guiding PCF.

2.9 Fabrication of photonic crystal fiber

From the invention of the first PCF in 1996, researchers have wanted to study various PCF designs. PCFs have several structural parameters that one can tailor with ease, thus they provide greater design flexibility and exhibit different physical properties from conventional optical fibers. The basic parameters of this type of fiber are hole size, hole position, pitch (center-to-center hole position), core diameter, and number of air-hole rings [24, 25]. The combined effects of surface tension and hole pressure during fiber drawing make it very difficult to control the shape of the holes in the fiber structure. Changes in the size and shape of the air holes can be made during fiber drawing, thus causing deviations between the fiber's profile and the drawing preform's profile [26]. To fabricate PCFs with hexagonal lattices and circular holes in the cladding, the stack and draw method has been adopted. The stacking method is the most common method for preparing a PCF preform, but it often has minor defects that affect the transmission of light [27]. Here, the preforms are stacked using circular silica capillaries. PCF preforms prepared by traditional mechanical drilling methods used to have the problem of inaccurate hole positioning [28]. Today, drilling methods based on computerized numerical control procedures are used to prepare PCF preforms with perfect hole positions and complex structures.

2.10 Material selection for PCF modeling

In general, most studies of PCFs have been based on silica glass material. Soft glass, especially chalcogenide glass, has excellent characteristics such as large nonlinearity and infrared transmission. However, the difficulty of preparing PCFs limits the usage of chalcogenide PCF's. In matrix selection, the high refractive index of chalcogenide glass enables it to achieve ultrahigh nonlinearity with a relatively short length [29]. In addition, chalcogenide glass has a lower glass transition temperature than other materials, which is beneficial to PCF manufacture. Liu *et al* found that the performance of a three-core PCF based on ZnTe (tellurite) glass was better than that of silica glass [30]. Apart from changing the geometry of the PCF, an alternative method to control its transmission and polarization properties is to fill the air holes, either completely or selectively, with various liquids such as CS_2, nitrobenzene, chloroform, water, ethanol, polymers, or liquid crystals [31]. Liquid-core PCFs (LCPCFs) consisting of liquid cores with numerous periodically spaced air holes in the cladding region have attracted a great deal of attention. A schematic diagram of an LCPCF is shown in figure 2.11. The possibility of filling PCF with liquids offers an enormous increase in the nonlinearity value of the fiber, adjustable dispersion, and endless single-mode operation. For instance, Voronin *et al* demonstrated that a hollow PCF filled with a highly nonlinear liquid can support single-mode guiding at wavelengths longer than 600nm in a 4 μm diameter liquid core [32]. The nonlinear response of such a fiber has been shown to drastically differ from the typical nonlinear response of a silica PCF; the authors of [33] showed that the strong inertia of optical nonlinearity of the liquid filling the fiber core translates into a pulse-width-dependent red shift of the spectrally broadened fiber output. In 2006, Wolinski *et al*

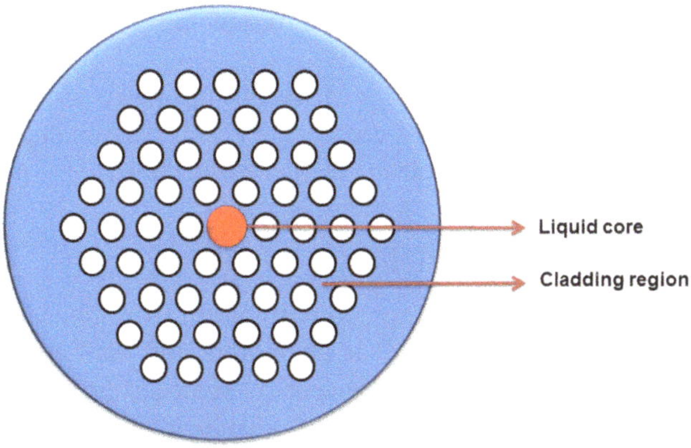

Figure 2.11. Schematic diagram of a liquid-core PCF with an air-hole diameter of d and pitch of Λ. The core is filled with liquid.

reported the effect of temperature and external electrical field on a PCF filled with a prototype nematic liquid crystal characterized by extremely low birefringence. They controlled switching between different guiding mechanisms by introducing the liquid-crystal material into the micro holes of the liquid-filled PCF. This development offers great potential for fiber-optic sensing and optical processing applications. Recently, Olausson *et al* fabricated an all-spliced fiber laser cavity with a single-mode output using a liquid-crystal PBGF which was electrically tunable from 1040 nm to 1065 nm [34]. This liquid-crystal PBGF device induced a bandgap in the cavity, which was tuned by an electric field and used to shift the laser wavelength by 25 nm.

2.11 Advantages

Due to its design flexibility, the properties of a PCF can be manipulated by adjusting the air-hole diameter or lattice pitch and using different materials for the cladding. The engineering of PCF properties has a wide range of advantages compared with the engineering of conventional fibers, for example:

1. The index difference between the core and cladding indices is achieved by adjusting the air-hole diameter (d) and lattice pitch (Λ) parameters. Similarly, different index variations can be induced by filling the core or air holes with polarizable liquids.
2. Because of the large index difference compared to that of conventional optical fiber, PCFs can create better light confinement in the core, allowing light guidance to take place in a core as small as 2 μm in diameter.
3. PCFs can be modeled with a much smaller effective area than those of standard fibers by varying the geometrical parameters (d and Λ), which leads to a high nonlinear coefficient (γ).
4. The group velocity dispersion (GVD) coefficient (β_2) of the PCF can be varied by altering the structural parameters (d and Λ). This is known as dispersion engineering.

5. A PCF can be designed with two ZDWs in the optical spectrum, whereas ordinary fibers only have one ZDW.
6. Because the relative contributions of various nonlinear processes are dependent on the dispersion profile of the fiber, PCF dispersion engineering may be utilized to influence the spectrum of the output pulse.

2.12 Pulse compression in PCF

For many nonlinear optics applications, in particular, for nonlinear pulse compression, the use of a free space geometry is unsuitable and confining geometries are necessary to enhance the effect of the nonlinearity. Hence, soliton-based transmission systems have become leading candidates for ultrahigh-speed long-haul light-wave transmission links, since they offer the possibility of a dynamic balance between GVD and SPM, which are the two effects that severely limit the performance of nonsoliton systems [35]. Because the dispersion and nonlinear properties of PCF can be tailored to the desired level by selecting an appropriate air-hole size and pitch, the invention of PCF has provided a new tool for pulse compression through the generation of solitons. To date, soliton transmission experiments have successfully achieved 160 Gbits/s over a distance of 10 000 km [36]. It has been shown through both experiments and numerical simulations that optical pulses can be compressed to very short widths using PCF with a propagation length of a few centimeters.

Based on soliton propagation in PCF, optical pulses generated by semiconductor lasers can be compressed using techniques such as soliton effect compression, [37–39] adiabatic pulse compression, [40–44] and self-similar compression. For instance, Tse *et al* numerically examined femtosecond soliton compression in PCF at a wavelength of 1.55 μm and revealed the optimum PCF parameters and improved soliton compression lengths [45]. In 2006, four-layer geometry was used to achieve a compression factor of 10 in a pulse with an initial full width at half maximum (FWHM) duration of 3 ps in a tapered fiber of 28 m long [46]. Prior to the invention of PCF, pulse compression had not been demonstrated at wavelengths shorter than $\approx$1.3 μm because of the requirement for anomalous GVD, which is not possible in conventional single-mode optical fiber below this wavelength. However, considerable effects have been directed toward the generation of ultrashort optical pulses at shorter wavelengths since the invention of PCFs. In this context, Travers *et al* compressed sub-50 fs pulses by a factor of 15 at a wavelength of 1.06 μm, indicating that tapered PCF may achieve large pulse-compression ratios [47]. Furthermore, an optimized 3.7 fs pulse was generated by compressing an initial ultrashort pulse width of 100 fs and an energy of 0.5 nJ centered at 800 nm [48]. Similarly, at a central wavelength of 1070 nm, a numerical model of pulse compression was proven by solving the generalized nonlinear Schrödinger equation (GNLSE) for the generation of 2 fs pulses with compression ratios of up to 50 [49]. Fu *et al* demonstrated the compression of low-power 6 ps pulses to 420 fs at 1550 nm in a compact As_2Se_3 SMF with high nonlinearity and positive normal dispersion in combination with a customized chirped fiber Bragg grating [50]. A detailed description of each pulse-compression scheme is given in the following chapters.

References

[1] Knight J C, Birks T A, Russell P St J and Atkin D M 1996 All-silica single-mode optical fiber with photonic crystal cladding *Opt. Lett.* **21** 1547–9
[2] Russel P S J 2003 Photonic crystal fibers *Science* **299** 358–62
[3] Yang S, Zhang Y, Peng X, Lu Y, Xie S, Li J, Chen W, Jiang Z, Peng J and Li H 2006 Theoretical study and experimental fabrication of high negative dispersion photonic crystal fiber with large area mode field *Opt. Express* **14** 3015–23
[4] Russell P St J 2006 Photonic-crystal fibers *J. Lightwave Technol.* **24** 4729–49
[5] Saitoh K, Koshiba M, Hasegawa T and Sasaoka E 2003 Chromatic dispersion control in photonic crystal fibers: application to ultra-flattened dispersion *Opt. Express* **11** 843–52
[6] Poli F, Cucinotta A, Selleri S and Bouk A H 2004 Tailoring of flattened dispersion in highly nonlinear photonic crystal fibers *IEEE Photonics Technol. Lett.* **16** 1065–7
[7] Ortigosa-Blanch A, Knight J C, Wadsworth W J, Arriaga J, Mangan B J, Birks T A and Russell P St J 2000 Highly birefringent photonic crystal fibers *Opt. Lett.* **25** 1325–7
[8] Matsui T, Zhou J, Nakajima K and Sankawa I 2005 Dispersion-flattened photonic crystal fiber with large effective area and low confinement loss *J. Lightwave Technol.* **23** 4178–83
[9] Hu D J J, Shum P P, Lu C and Ren G 2009 Dispersion-flattened polarization-maintaining photonic crystal fiber for nonlinear applications *Opt. Commun.* **282** 4072–6
[10] Birks T A, Knight J C and Russell P St J 1997 Endlessly single-mode photonic crystal fiber *Opt. Lett.* **22** 961–3
[11] Le T, Hofer M, Cheng Z, Stingl A, Darmo J, Kelly D P and Unterrainer K 2008 Advances in fiber delivery of ultrashort pulses at 800 nm *Solid State Lasers XVII: Technology and Devices* (Bellingham, WA: SPIE) 613–22
[12] Zhang X, Liu Z, Gui Y, Gan H, Guan Y, He L, Wang X, Shen X and Dai S 2021 Characteristics and preparation of a polarization beam splitter based on a chalcogenide dual-core photonic crystal fiber *Opt. Express* **29** 39601–10
[13] Petrovich M N, Poletti F, Van Brakel A and Richardson D J 2008 Robustly single mode hollow core photonic bandgap fiber *Opt. Express* **16** 4337–46
[14] Passaro D, Foroni M, Poli F, Cucinotta A, Selleri S, Laegsgaard J and Bjarklev A O 2008 All-silica hollow-core microstructured Bragg fibers for biosensor application *IEEE Sens. J.* **8** 1280–6
[15] Saitoh K and Koshiba M 2003 Leakage loss and group velocity dispersion in air-core photonic bandgap fibers *Opt. Express* **11** 3100–9
[16] Gauvreau B, Hassani A, Fehri M F, Kabashin A and Skorobogatiy M 2007 Photonic bandgap fiber-based surface plasmon resonance sensors *Opt. Express* **15** 11413–26
[17] Knight J C 2007 Photonic crystal fibers and fiber lasers *J. Opt. Soc. Am.* B **24** 1661–8
[18] Benabid F, Bouwmans G, Knight J C, Russell P St J and Couny F 2004 Ultrahigh efficiency laser wavelength conversion in a gas-filled hollow core photonic crystal fiber by pure stimulated rotational Raman scattering in molecular hydrogen *Phys. Rev. Lett.* **93** 123903
[19] Couny F, Benabid F and Light P S 2007 Subwatt threshold CW Raman fiber-gas laser based on H2-filled hollow-core photonic crystal fiber *Phys. Rev. Lett.* **99** 143903
[20] Dasgupta S, Pal B P and Shenoy M R 2007 Nonlinear spectral broadening in solid-core Bragg fibers *J. Lightwave Technol.* **25** 2475–81
[21] de Matos C J S 2008 Modeling long-pass filters based on fundamental-mode cutoff in photonic crystal fibers *IEEE Photonics Technol. Lett.* **21** 112–4

[22] Michaille L, Bennett C R, Taylor D M and Shepherd T J 2009 Multicore photonic crystal fiber lasers for high power/energy applications *IEEE J. Sel. Top. Quantum Electron.* **15** 328–36
[23] Koizumi K, Yoshida M and Nakazawa M 2010 A 10-GHz optoelectronic oscillator at 1100 nm using a single-mode VCSEL and a photonic crystal fiber *IEEE Photonics Technol. Lett.* **22** 293–5
[24] Varshney S K, Saitoh K and Koshiba M 2005 A novel design for dispersion compensating photonic crystal fiber Raman amplifier *IEEE Photonics Technol. Lett.* **17** 2062–4
[25] Mortensen N A, Nielsen M D, Folkenberg J R, Petersson A and Simonsen H R 2003 Improved large-mode-area endlessly single-mode photonic crystal fibers *Opt. Lett.* **28** 393–5
[26] Kuhlmey B T, Luan F, Fu L, Yeom D-I, Eggleton B J, Wang A and Knight J C 2008 Experimental reconstruction of bands in solid core photonic bandgap fibres using acoustic gratings *Opt. Express* **16** 13845–56
[27] Brilland L *et al* 2009 Recent progress on the realization of chalcogenides photonic crystal fibers *Optical Components and Materials VI* (Bellingham, WA: SPIE)
[28] El-Amraoui M *et al* 2010 Microstructured chalcogenide optical fibers from as 2 s 3 glass: towards new IR broadband sources *Opt. Express* **18** 26655–65
[29] Eggleton B J 2010 Chalcogenide photonics: fabrication, devices and applications introduction *Opt. Express* **18** 26632–4
[30] Liu S, Li S-G, Yin G-B, Feng R-P and Wang X-Y 2012 A novel polarization splitter in ZnTe tellurite glass three-core photonic crystal fiber *Opt. Commun.* **285** 1097–102
[31] Conti C, Schmidt M A, Russell P St J and Biancalana F 2010 Highly noninstantaneous solitons in liquid-core photonic crystal fibers *Phys. Rev. Lett.* **105** 263902
[32] Voronin A A, Mitrokhin V P, Ivanov A A, Fedotov A B, Sidorov-Biryukov D A, Beloglazov V I, Alfimov M V, Ludvigsen H and Zheltikov A M 2010 Understanding the nonlinear-optical response of a liquid-core photonic-crystal fiber *Laser Phys. Lett.* **7** 46–9
[33] Wolinski T R, Szaniawska K, Ertman S, Lesiak P, Domanski A W, Dabrowski R, Nowinowski-Kruszelnicki E and Wojcik J 2006 Influence of temperature and electrical fields on propagation properties of photonic liquid-crystal fibres *Meas. Sci. Technol.* **17** 985
[34] Olausson C B, Scolari L, Wei L, Noordegraaf D, Weirich J, Alkeskjold T T, Hansen K P and Bjarklev A 2010 Electrically tunable Yb-doped fiber laser based on a liquid crystal photonic bandgap fiber device *Opt. Express* **18** 8229–38
[35] Agrawal G P 2010 *Fiber-optic Communication Systems* 4 (New York: Wiley)
[36] Nakazawa M, Kubota H, Suzuki K, Yamada E and Sahara A 2000 Ultrahigh-speed long-distance TDM and WDM soliton transmission technologies *IEEE J. Sel. Top. Quantum Electron.* **6** 363–96
[37] Mollenauer L F, Stolen R H and Gordon J P 1980 Experimental observation of picosecond pulse narrowing and solitons in optical fibers *Phys. Rev. Lett.* **45** 1095
[38] Aamer Ahmed K, Chan K C and Liu H-F 1995 Femtosecond pulse generation from semiconductor lasers using the soliton-effect compression technique *IEEE J. Sel. Top. Quantum Electron.* **1** 592–600
[39] Chan K C and Liu H F 1995 Short pulse generation by higher order soliton-effect compression: effects of optical fiber characteristics *IEEE J. Quantum Electron.* **31** 2226–35
[40] Chernikov S V and Mamyshev P V 1991 Femtosecond soliton propagation in fibers with slowly decreasing dispersion *J. Opt. Soc. Am.* B **8** 1633–41
[41] Chernikov S V, Dianov E M, Richardson D J and Payne D N 1993 Soliton pulse compression in dispersion-decreasing fiber *Opt. Lett.* **18** 476–8

[42] Pelusi M D and Liu H-F 1997 Higher order soliton pulse compression in dispersion-decreasing optical fibers *IEEE J. Quantum Electron.* **33** 1430–9

[43] Tamura K R and Nakazawa H 1999 Femtosecond soliton generation over a 32-nm wavelength range using a dispersion-flattened dispersion-decreasing fiber *IEEE Photonics Technol. Lett.* **11** 319–21

[44] Chan K-T and Cao W-H 2000 Enhanced compression of fundamental solitons in dispersion decreasing fibers due to the combined effects of negative third-order dispersion and Raman self-scattering *Opt. Commun.* **184** 463–74

[45] Ming-Leung V T, Horak P, Poletti F and Richardson D J 2008 Designing tapered holey fibers for soliton compression IEEE *J. Quantum Electron.* **44** 192–8

[46] Hu J, Marks B S, Menyuk C R, Kim J, Carruthers T F, Wright B M, Taunay T F and Joseph Friebele E 2006 Pulse compression using a tapered microstructure optical fiber *Opt. Express* **14** 4026–36

[47] Travers J C, Stone J M, Rulkov A B, Cumberland B A, George A K, Popov S V, Knight J C and Taylor J R 2007 Optical pulse compression in dispersion decreasing photonic crystal fiber *Opt. Express* **15** 13203–11

[48] Tognetti M V and Crespo H M 2007 Sub-two-cycle soliton-effect pulse compression at 800 nm in photonic crystal fibers *J. Opt. Soc. Am.* B **24** 1410–5

[49] Voronin A A and Zheltikov A M 2008 Soliton-number analysis of soliton-effect pulse compression to single-cycle pulse widths *Phys. Rev.* A **78** 063834

[50] Littler I C M, Fu L B, Mägi E C, Pudo D and Eggleton B J 2006 Widely tunable, acousto-optic resonances in Chalcogenide As2Se3 fiber *Opt. Express* **14** 8088–95

[51] Poletti F, Petrovich M N and Richardson D J 2013 Hollow-core photonic bandgap fibers: technology and applications *Nanophotonics* **2** 315–40

[52] Van Le Hieu, Hoang Van Thuy and Stępniewski Grzegorz *et al* 2021 Low pump power coherent supercontinuum generation in heavy metal oxide solid-core photonic crystal fibers infiltrated with carbon tetrachloride covering 930–2500 nm *Opt. Express* **29** 39586–600

IOP Publishing

Fiber Optic Pulse Compression

Numerical techniques and applications with MATLAB®

R Vasantha Jayakantha Raja and A Esther Lidiya

Chapter 3

Theory and modeling of photonic crystal fiber

It is necessary to perform numerical simulations based on well-grounded physical models in order to propose new ideas. In this chapter, the existing numerical method used to solve the nonlinear Schrödinger equation (NLSE) is reviewed and the numerical models used to calculate photonic crystal fiber (PCF) parameters are discussed in detail.

3.1 Numerical methods

In general, the propagation of an ultrashort pulse and broadband radiation in PCF can be studied by solving the NLSE. The NLSE has analytic solutions only in some cases, i.e. those in which mechanisms such as self-steepening, Raman delayed response, and higher-order dispersion can be neglected [1]. However, the higher-order nonlinear effects and dispersions are essential for a deep understanding of spectral broadening mechanisms. In addition, in order to investigate pulse propagation, three PCF parameters, namely, loss α, the dispersion coefficients β_n, and the nonlinear parameter γ must be known. All these PCF parameters can be determined by calculating the effective refractive-index (n_{eff}) value of the PCF. The value of n_{eff} strongly depends on the geometry of the PCF cross section, in particular, on the size and periodicity of the holes [2]. Since the PCF structure is complex in nature, numerical simulation techniques are generally preferred for the calculation of the PCF parameters. These numerical simulation techniques are required in order to discuss the temporal and spectral characteristics of the supercontinuum (SC) and pulse compression, and to interpret the underlying physics of ultrashort and broad-spectrum processes. Based on this context, this chapter provides an efficient numerical procedure for calculating the PCF parameters and solving the NLSE. Even though many techniques have been proposed to solve the guidance properties of PCFs, two primary well-proven techniques for modeling PCFs are the fully vectorial effective index method (FVEIM) and the finite-element method (FEM), which are outlined in this chapter. In addition, the use of the split-step Fourier

doi:10.1088/978-0-7503-2686-5ch3

method (SSFM) to solve the NLSE is described in detail. Numerical studies of the complicated structure of PCF employ a wide variety of techniques, including the scalar effective index method (SEIM) [3–5], the beam propagation method (BPM) [6, 7], the finite difference method [8, 9], the FEM [10, 11], the plane-wave expansion method [12, 13], and the multipole method [14, 15]. Among the many numerical methods, the SEIM scheme requires only small amounts of computation time and computer memory. This method is the easiest and has been successfully applied to analyze the propagation properties of PCF [4]. While the SEIM is not valid for large hole diameters, the accuracy of this method is adequate for the investigation of small hole diameters. Based on the above, Li *et al* proposed the use of the FVEIM for the calculation of the effective refractive index. They proved that the FVEIM model gives more accurate results than the SEIM model [16]. Using this method, they calculated the dispersion, and the results were in agreement with experimental data and other numerical simulations.

3.2 The fully vectorial effective index method

The FVEIM is a numerical method that provides a fully vectorial analysis of the propagation modes of electromagnetic fields in PCF [2]. The PCF selected was an index-guiding type with a solid silica core surrounded by a hexagonal array of air holes running along its length. PCF is characterized by the pitch of its air holes and their diameter. To find the effective refractive index, the equivalent step-index fiber is considered, which has a cladding refractive index of $n_{\mathrm{FSM,cl}}$ and a core index of n_s. A schematic diagram of the PCF and the equivalent step-index fiber is shown in figure 3.1. The FVEIM analysis begins with Maxwell's equations. Since the refractive-index profiles of fibers are cylindrically symmetric, it is convenient to use the cylindrical polar coordinate system. The homogeneous vector wave

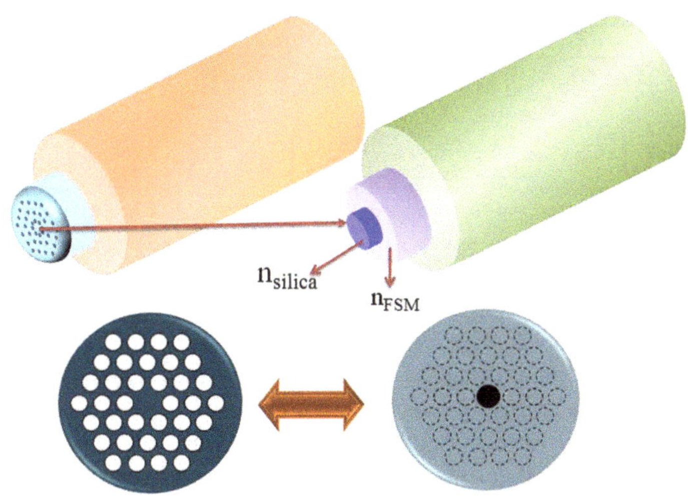

Figure 3.1. Schematic diagram of PCF and its equivalent step-index fiber.

equations for the electrical field $\boldsymbol{E}$ and magnetic field $\boldsymbol{H}$ can be rewritten in cylindrical polar coordinates as [2, 17]

$$\frac{\partial^2\psi}{\partial r^2} + \frac{1}{r}\frac{\partial\psi}{\partial r} + \frac{1}{r^2}\frac{\partial^2\psi}{\partial\phi^2} + (k^2n^2 - \beta^2)\psi = 0, \tag{3.1}$$

where n is the refractive index of the material, β is the propagation constant, k is the wave vector, and ψ can represent either E_z or H_z. Using the separation of variables, one can easily solve this equation. The solution to the cylindrical wave equation consists of the regular and modified Bessel functions in terms of J, Y and I, K, respectively. It should be noted that several Bessel functions can be mathematically valid for this equation (3.1). However, based on the physical nature of the problem, only a few of these solutions can be selected as valid. With these Bessel functions as the solutions, the z-components of the fields within the core and cladding can be obtained, as follows:

$$\begin{aligned} E_z(r,\omega) &= AJ_l(hr)exp[i(\omega t + l\phi - \beta z)] \\ H_z(r,\omega) &= BJ_l(hr)exp[i(\omega t + l\phi - \beta z)] \\ &\qquad\qquad\qquad 0 < r < \rho \\ E_z(r,\omega) &= CK_l(\gamma r)exp[i(\omega t + l\phi - \beta z)] \\ H_z(r,\omega) &= DK_l(\gamma r)exp[i(\omega t + l\phi - \beta z)] \\ &\qquad\qquad\qquad \rho < r < b, \end{aligned} \tag{3.2}$$

where A, B, C, and D are constants, $h(\omega)^2 = k^2n_s^2 - \beta(\omega)^2$ and $\gamma(\omega)^2 = \beta(\omega)^2 - k^2n_{\mathrm{FSM}}^2$. By assuming the continuity of the field and its radial derivative across each fiber boundary, a set of four equations with four unknowns is generated. This set of equations is used to determine the eigenvalue equation which helps us to find the propagation constants of the different fiber modes. Using the above consideration, the eigenvalue equation can be obtained, as follows:

$$\begin{aligned} &\left(\frac{J_l'(h\rho)}{h\rho J_l(h\rho)} + \frac{K_l'(\gamma\rho)}{\gamma\rho K_l(\gamma\rho)}\right)\left(\frac{n_s^2 J_l'(h\rho)}{q\rho J_l(h\rho)} + \frac{n_{\mathrm{FSM}}^2 K_l'(\gamma\rho)}{\gamma\rho K_l(\gamma\rho)}\right) \\ &\qquad = l^2\left[\left(\frac{1}{h\rho}\right)^2 + \left(\frac{1}{\gamma\rho}\right)^2\right]^2\left(\frac{\beta}{k}\right)^2, \end{aligned} \tag{3.3}$$

where radius of the core is given by $\rho = 0.64\Lambda$ [18, 19]. The solution to this eigenvalue equation obtained by setting $l = 1$ gives the propagation constant for the HE_{11} mode. From this propagation constant, one can easily obtain the effective refractive index of the PCF. However, it should be noted from the above equation that the effective cladding index n_{FSM} is still unknown. In order to calculate n_{FSM}, we split the photonic crystal into hexagonal unit cells so that the air holes are in the centers of the cells. To further simplify the calculations, the hexagonal unit cell is treated as a circular cell, as shown in figure 3.2. The Bessel solution for the core and cladding regions of the unit cell can now be obtained as follows:

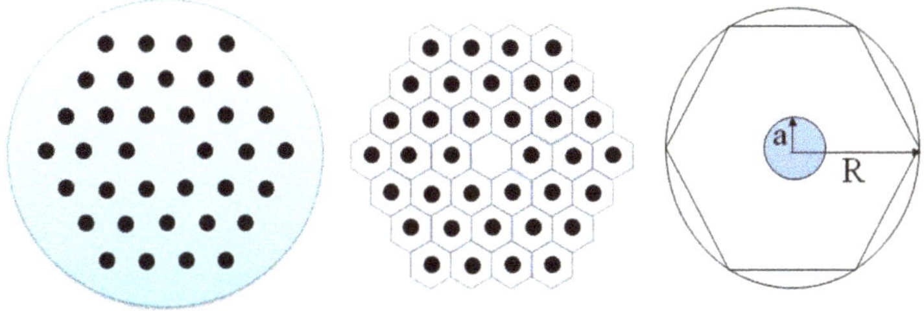

Figure 3.2. The formation of air-hole unit cells used to calculate the effective cladding refractive index.

$$\begin{aligned} E_z(r,\omega) &= I_l[w(\omega)r] \\ H_z(r,\omega) &= I_l[w(\omega)r] \\ &0 < r < a \\ E_z(r,\omega) &= J_l[q(\omega)r]Y_l[q(\omega)R] - Y_l[q(\omega)r]J_l[q(\omega)R] \\ H_z(r,\omega) &= J_l[q(\omega)r]Y_l[q(\omega)R] - Y_l[q(\omega)r]J_l[q(\omega)R] \\ &\rho < r < R, \end{aligned} \tag{3.4}$$

where $w(\omega)^2 = \beta(\omega)^2 - k^2n_0^2$, n_0 is the refractive index of air, $q(\omega)^2 = k^2n_s^2 - \beta(\omega)^2$, and the outer circle radius $R = \Lambda\sqrt{\frac{\sqrt{3}}{2\pi}}$ [2, 20, 21]. I, J, and Y are Bessel functions. Boundary conditions are applied to the wave equation and the characteristic equation for the dispersion relations to find the effective cladding index, which is obtained from

$$\begin{aligned} &\left(\frac{P_l'(qa)}{qaP_l(qa)} + \frac{I_l'(wa)}{waI_l(wa)}\right)\left(\frac{P_l'(qa)}{qaP_l(qa)} + \frac{I_l'(wa)}{waI_l(wa)}\right) \\ &= l^2\left[\left(\frac{1}{qa}\right)^2 + \left(\frac{1}{wa}\right)^2\right]^2\left(\frac{\beta_{\mathrm{FSM}}}{k}\right)^2, \end{aligned} \tag{3.5}$$

where

$$P_l[q(\omega)r] = J_l[q(\omega)r]Y_l[q(\omega)R] - Y_l[q(\omega)r]J_l[q(\omega)R].$$

The value of this propagation constant is dependent on the parameters of the PCF. Given the propagation constant n_{FSM} associated with the space-filling mode, the effective refractive index of the cladding of the equivalent fiber can be deduced from

$$n_{\mathrm{FSM,\,cl}} = 2\pi\beta/\lambda. \tag{3.6}$$

The determination of the effective cladding index and the core index using the combination of equations (3.5) and (3.3) is called the FVEIM. Using the FVEIM, the effective cladding index n_{FSM} can be obtained as a function of λ by numerical simulation for various diameters of the air hole and pitch, as shown in figures 3.3(a) and (b). The changes of n_{eff} were obtained as a function of λ for various air-hole diameters, namely 0.4, 0.6 0.8, 1.0, and 1.2 μm, and values of Λ, namely 1.6, 1.8, 2.0,

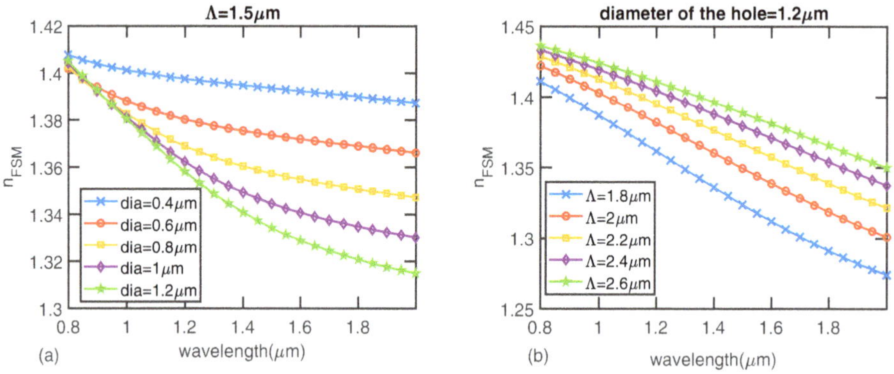

Figure 3.3. Variation of effective cladding index as a function of wavelength calculated by FVEIM (a) for air-hole diameters of 0.4, 0.6, 0.8, 1.0, and 1.2 μm with a fixed value of $\Lambda = 1.5$ μm, (b) for various pitch values with a fixed air-hole diameter of 1.2 μm.

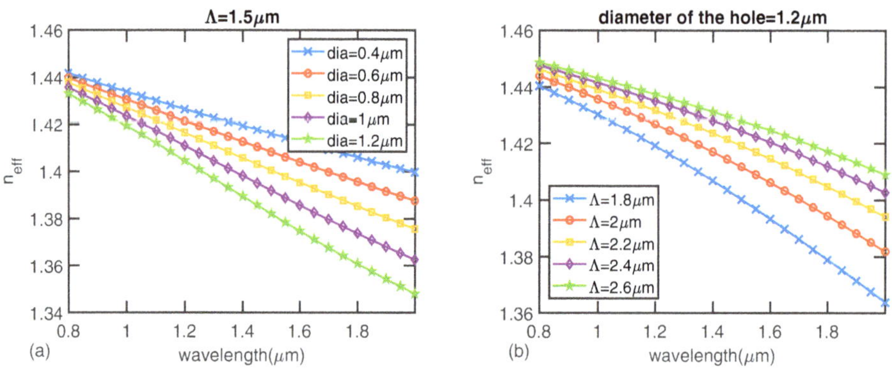

Figure 3.4. Effective indexes of the fundamental mode n_{eff} as a function of wavelength (a) calculated for air-hole diameters of 0.4, 0.6, 0.8, 1.0, and 1.2 μm by fixing a pitch value of $\Lambda = 1.5$ μm, (b) for pitch values of 1.8, 2.0, 2.2, 2.4, and 2.6 μm with an air-hole diameter of 1.2 μm.

2.2, 2.4, and 2.6 μm, as shown in figures 3.4(a) and (b). As the air filling factor grows as a result of changing the size of the hole, n_{FSM} shifts toward lower values. Hence, it is obvious that the values of n_{eff} and n_{FSM} decrease as the wavelength increases.

3.3 Group velocity dispersion (GVD)

Chromatic dispersion is very important among the effects that cause degradation in fiber that arise from the frequency dependence of the refractive index $n(\omega)$. Since the velocity of light is determined by $c/n(\omega)$, the different spectral components associated with the pulse travel at different speeds. In general, the refractive index of a material depends on the wavelength of the electromagnetic wave interacting with the material. This dependence is known as material dispersion and it can be represented using the Sellmeier approximation [1]:

$$n(\lambda) = \sqrt{1 + \sum_{k=1}^{m} \frac{B_k \lambda_k^2}{\lambda_k^2 - \lambda^2}}, \tag{3.7}$$

where B_k is the magnitude of the kth resonance of the material located at the operating wavelength. For silica fibers, the refractive index can be closely approximated using the following values for B_k and λ_k: $B_1 = 0.696\,1663$, $B_2 = 0.407\,9426$, $B_3 = 0.897\,4794$ and $\lambda_1 = 0.068\,4043\ \mu\text{m}$, $\lambda_2 = 0.116\,2414\ \mu\text{m}$ and $\lambda_3 = 9.896\,161\ \mu\text{m}$. Dispersion-induced spectral broadening would be very important, even without nonlinearity. The effects of dispersion can be accounted for by expanding the mode-propagation constant in a Taylor series around the central frequency ω_0, as follows:

$$\beta(\omega) = n(\omega)\frac{\omega}{c} = \beta_0 + \beta_1(\omega - \omega_0) + \frac{1}{2}\beta_2(\omega - \omega_0)^2 + \cdots, \tag{3.8}$$

where $\beta_m = \left(\frac{d^m\beta}{d\omega^m}\right)_{\omega=\omega_0}$. Thus, it is easy to get the first- and second-order derivatives from (3.8):

$$\beta_1 = \frac{1}{v_g} = \frac{n_g}{c} = \frac{1}{c}\left(n + \omega\frac{dn}{d\omega}\right) \tag{3.9}$$

$$\beta_2 = \frac{1}{c}\left(2\frac{dn}{d\omega} + \omega\frac{d^2n}{d\omega^2}\right), \tag{3.10}$$

where n_g is the group index and v_g is the group velocity. The wavelength at which $\beta_2 = 0$ is called the zero-dispersion wavelength, λ_D. However, there is still dispersion at the wavelength λ_D, and higher-order dispersion will be considered in this case. Another parameter involved in the dispersion of fiber is the dispersion parameter D. The relationship between D and β_1, β_2 is defined as follows:

$$D = \frac{d\beta_1}{d\lambda} = -\frac{2\pi c}{\lambda^2}\beta_2 = -\frac{\lambda}{c}\frac{d^2n}{d\lambda^2}. \tag{3.11}$$

From equation (3.11), we can see that D has the opposite sign to that of β_2. Figures 3.5(a) and (b) show the measured variation of the dispersion parameter D versus wavelength for a solid-silica-core PCF. In the regime in which wavelength $\lambda < \lambda_D$, $\beta_2 > 0$ (or $D < 0$), the fiber is said to exhibit normal dispersion. In the normal dispersion regime, the high-frequency components of an optical signal travel slower than its low-frequency components. By contrast, in the regime in which wavelength $\lambda > > \lambda_D$, $\beta_2 < 0$ (or $D > 0$), fiber is said to exhibit anomalous dispersion. In the anomalous dispersion regime, the high-frequency components of a signal travel faster than its low-frequency components. From figures 3.5(a) and (b), it can be observed that it is possible to shift the zero-dispersion wavelength of the fiber to any value from 500 nm to beyond 1500 nm by altering the size of the air hole and pitch.

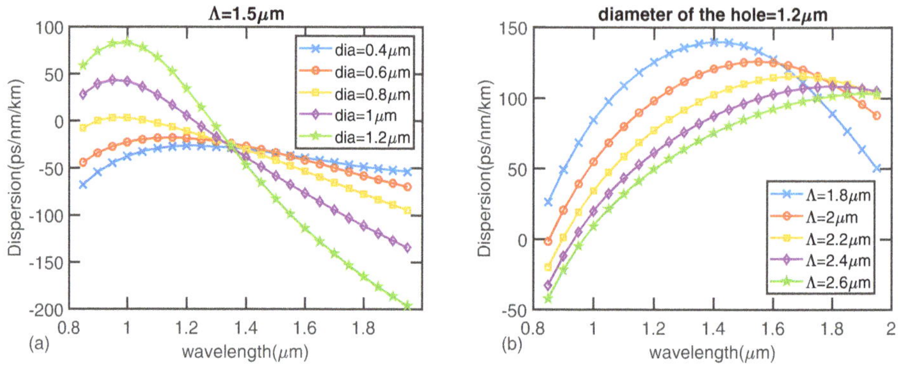

Figure 3.5. Group velocity dispersion of PCF, including the material dispersion of the fiber, as a function of wavelength calculated by FVEIM (a) for $\Lambda = 1.5\ \mu$m, (b) for various pitch values.

3.4 Mode parameters of PCF

The number of modes supported by a PCF depends on refractive indices of the core and cladding. The refractive index of the cladding can be adjusted by varying the d/Λ values. Using a similar approach to that used for conventional fiber, one can calculate the normalized parameter V as follows [22, 23]:

$$V_{\text{eff}} = \frac{2\pi}{\lambda}\rho\sqrt{n_s^2 - n_{\text{cl}}^2} = \sqrt{U_{\text{eff}}^2 + W_{\text{eff}}^2}, \tag{3.12}$$

where

$$U_{\text{eff}} = \frac{2\pi}{\lambda}\rho\sqrt{n_s^2 - n_{\text{eff}}^2}, \tag{3.13}$$

and

$$W_{\text{eff}} = \frac{2\pi}{\lambda}\rho\sqrt{n_{\text{eff}}^2 - n_{\text{cl}}^2}. \tag{3.14}$$

where n_s and n_{cl} are the refractive indices of the core and cladding, respectively, and ρ is the core radius. The normalized propagation constant 'b' is defined as

$$b_{\text{eff}} = \frac{(\beta/k_0)^2 - n_{\text{cl}}^2}{n_s^2 - n_{\text{cl}}^2}. \tag{3.15}$$

Using FVEIM, the U parameter, W parameter, and V parameter are calculated as a function of wavelength λ, as illustrated in figures 3.6, 3.7, and 3.8 for different values of the hole diameter to pitch ratio (d/Λ). For PCFs with a small relative air-hole diameter d/Λ, the V-parameter value is smaller than 2.405 for wavelengths between 400 nm and 1600 nm, thus proving that PCF can indeed be endlessly single mode. Such PCFs are utilized in SC experiments in which ultra-broadband coherent light can be generated in a single mode. However, large values of d/Λ result in multimode operation at the operating wavelength.

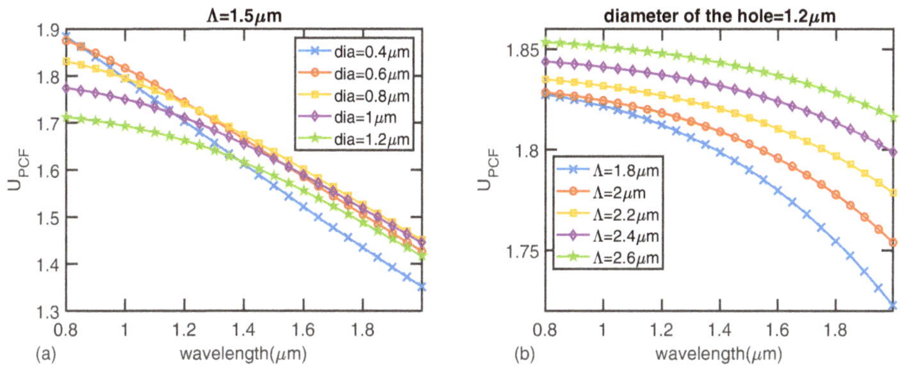

Figure 3.6. Variation of the *U* parameter of PCF as a function of wavelength (a) for a fixed value of $\Lambda = 1.5$ μm, (b) for different Λ values with a fixed air-hole diameter of 1.2 μm.

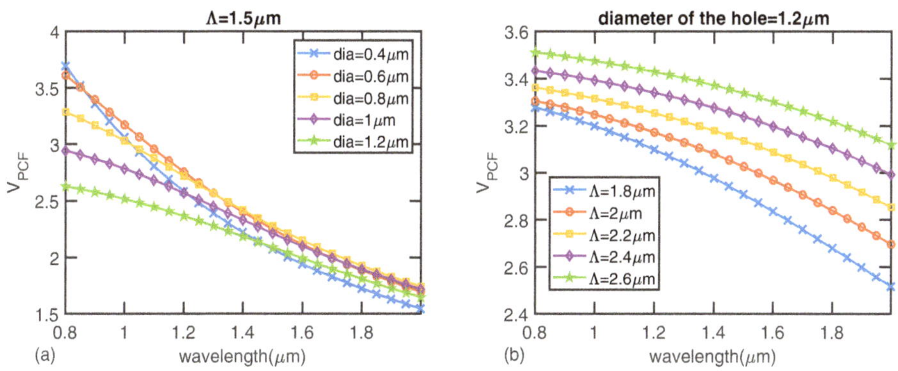

Figure 3.7. Variation of the *V* parameter as a function of wavelength (a) for different air-hole diameters and a Λ value of 1.5 μm, (b) for different Λ values with a fixed air-hole diameter of 1.2 μm.

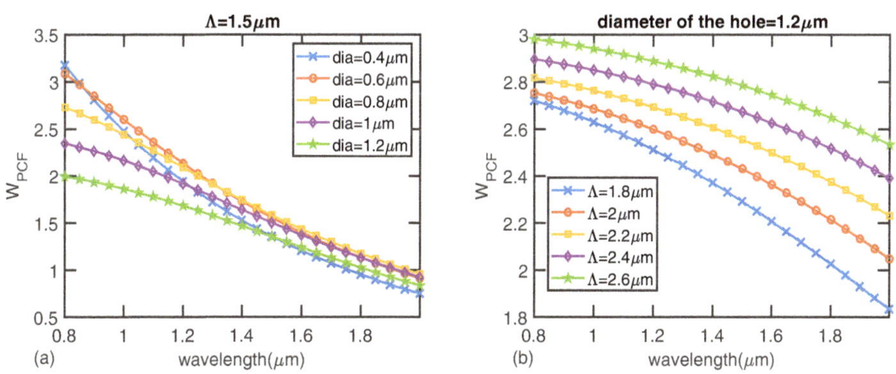

Figure 3.8. Variation of the *U* parameter of PCF as a function of wavelength (a) for a fixed value of $\Lambda = 1.5$ μm, (b) for different Λ values with a fixed air-hole diameter of 1.2 μm.

3.5 Linear properties of photonic crystal fiber

The spot size is a very important characteristic that determines various important parameters such as the splice loss, bend loss, launching effects, etc. The first spot size of the near field of a single-mode fiber is defined as [17, 24]

$$w_{\mathrm{PCFI}}^2 = \frac{2\int_0^\infty \psi^2(r) r^2 dr}{\int_0^\infty \psi^2(r) r dr}, \tag{3.16}$$

where $\psi(r)$ is the modal field distribution of the LP_{01} mode of the fiber. This is sometimes called the Petermann-I spot size and is related to the loss due to angular mismatch at a splice/joint between two fibers. In general, one has to evaluate the above integrals numerically, but for a step-index fiber, an analytical expression can be obtained, as follows:

$$w_{\mathrm{PCFI}} = \frac{2\rho}{\sqrt{3}}\left[\frac{J_0(U_{\mathrm{eff}})}{U_{\mathrm{eff}} J_1(U_{\mathrm{eff}})} + \frac{1}{2} + \frac{1}{W_{\mathrm{eff}}^2} - \frac{1}{U_{\mathrm{eff}}^2}\right], \tag{3.17}$$

where U_{eff} and W_{eff} are the normalized fiber parameters. The second spot size, w_{PCFII}, known as the Petermann-II spot size of the near field, is defined as

$$w_{\mathrm{PCFII}}^2 = \frac{2\int_0^\infty \psi^2(r) r dr}{\int_0^\infty \left(\frac{d\psi(r)}{dr}\right)^2 r dr}. \tag{3.18}$$

This spot size is related to the effective area of the core and to the nonlinearity of the PCF. w_{PCFII} can be written as

$$w_{\mathrm{PCFII}} = \sqrt{2}\rho \frac{J_1(U_{\mathrm{eff}})}{W_{\mathrm{eff}} J_0(U_{\mathrm{eff}})}. \tag{3.19}$$

Figures 3.9 and 3.10 show plots of spot size I and spot size II for different PCF designs in terms of λ. It can be predicted from these figures that the spot size should decrease as the size of the air hole increases and increase as the wavelength increases.

3.6 Nonlinear properties of photonic crystal fiber

The effective area is an important functional parameter that determines the optical performance of a PCF. It is a function of wavelength, core diameter, and the refractive-index difference between the core and cladding. The effective area A_{eff} is defined for the purposes of calculating nonlinear effects and characterizing fibers with standardized parameters. If one uses the Gaussian approximation for the mode, then

$$A_{\mathrm{eff}} = \pi w_{\mathrm{PCF}}^2, \tag{3.20}$$

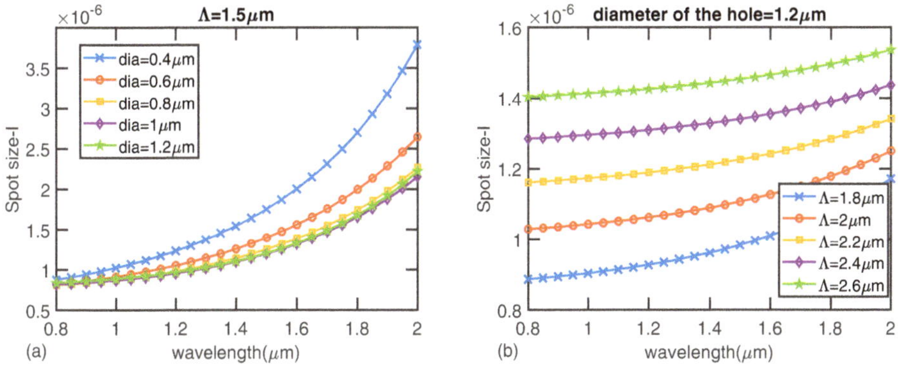

Figure 3.9. Spot size I of PCF as a function of wavelength calculated by FVEIM (a) for varying air-hole diameters of 0.4 to 1.2 μm in steps of 0.2 μm, (b) for varying Λ values with a fixed air-hole diameter of 1.2 μm.

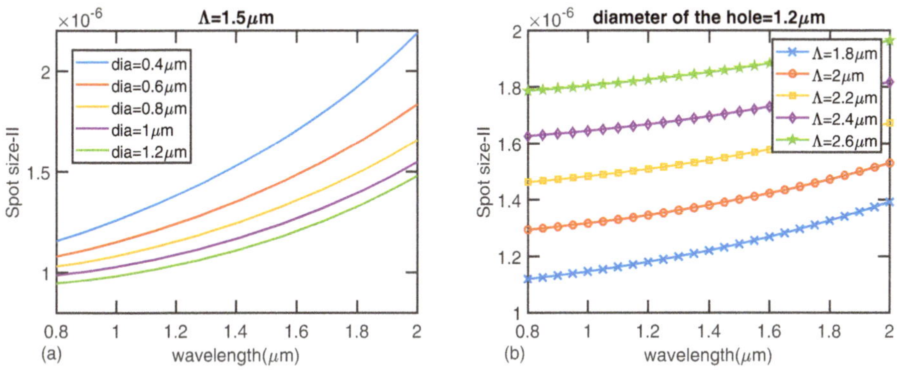

Figure 3.10. Variation of spot size II of PCF as a function of wavelength (a) for varying air-hole diameters from 0.4 to 1.2 μm in steps of 0.2 μm, (b) for varying Λ values with a fixed air-hole diameter of 1.2 μm.

where w_{PCF}^2 is the Gaussian spot size of the mode. Figure 3.11(a) shows the effective area calculated using the FVEIM as a function of wavelength for different air-hole diameters ranging from 0.4 μm to 1.2 μm in steps of 0.2 μm. Figure 3.11(b) depicts the dependence of the effective area for various values of Λ. The maximum effective area supported by a PCF depends strongly on the hole diameter d, pitch Λ, and wavelength λ. If the effective area is increased, then the influence of the intensity-dependent nonlinear effects is reduced. The small air holes in the cladding region of the PCF are introduced to obtain a large effective area and can also yield low bending loss, low dispersion, and low attenuation.

The nonlinear properties of the PCF depend mainly on the effective area A_{eff} and the nonlinear refractive index n_2 of the material. After calculating the effective area of the PCF, it is easy to calculate nonlinear parameter γ, as follows:

$$\gamma = \frac{n_2 \omega_0}{c A_{\text{eff}}}. \tag{3.21}$$

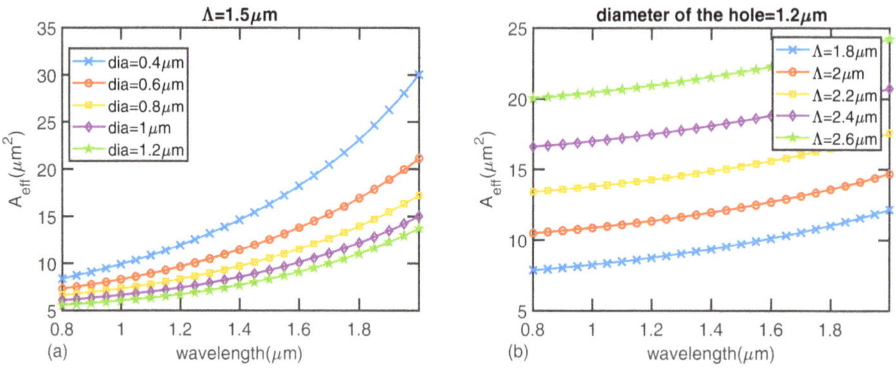

Figure 3.11. (a) Effective area as a function of wavelength for different hole sizes. The wavelength is scaled by Λ. The multipole method and the FVEIM are represented by solid and dashed lines, respectively. (b) The effective area of PCF as a function of the normalized wavelength by air-hole diameter for different d/Λ ratios.

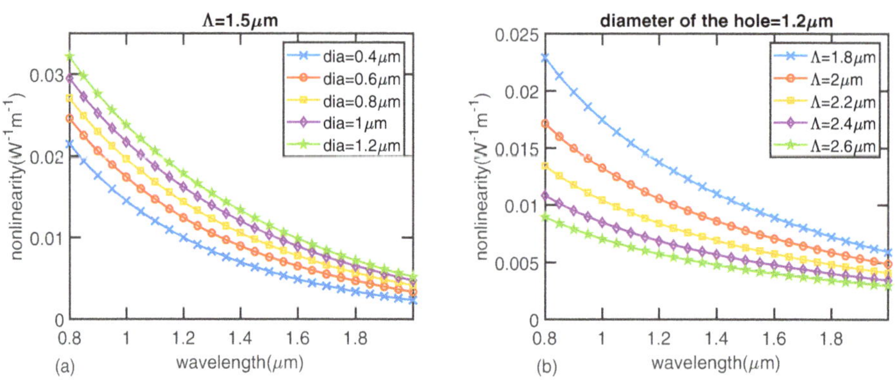

Figure 3.12. Nonlinear parameter as a function of wavelength (a) for different hole sizes, (b) for different Λ values. The multipole method and the FVEIM are represented by solid and dashed lines, respectively.

The n_2 value of silica is 3.0×10^{-20} m^2 W^{-1}. The nonlinear parameter of PCF was calculated for various air holes and is shown in figure 3.12(a). These calculations were repeated for different Λ values and the obtained results are shown in figure 3.12(b). From these plots it is clear that since the maximum effective area supported by a PCF depends on the hole diameter d and the wavelength λ, if the effective area is increased, then the influence of the intensity-dependent nonlinear effects is reduced.

3.7 Finite-element method

FEMs are a general class of numerical techniques for solving partial differential equations [25, 26]. They are well suited for problems that need to be solved for an irregular domain. In particular, this method is superior to other methods if the transverse structure exhibits a large refractive-index difference. The FEM allows the PCF cross section in the transverse x–y plane to be divided into a patchwork of triangular elements, which can be of different sizes, shapes, and refractive indices. In this way, any kind of PCF geometry can be accurately described. The boundaries of the triangles can be connected with the help of transition conditions. This scheme leads to a matrix eigenvalue system, which can be solved numerically. The FEM provides a full-vector analysis, which is necessary to model PCFs with large air holes and high index variations and to accurately predict their properties. Starting from Maxwell's equations, the basic equation for the FEM analysis is [25, 26]

$$\nabla \times ([\mu_r]^{-1}\nabla \times E) - k_0^2[\varepsilon_r]E = 0, \tag{3.22}$$

where $[\mu_r]$ and $[\varepsilon_r]$ are the relative magnetic permeability and the dielectric permittivity tensors, respectively. When applying full-vector FEM, the PCF cross-sectional domain is divided into subdomains consisting of triangular elements that can properly represent any refractive-index profiles. Dividing the PCF cross section into a number of triangular elements, using equation (3.22) we can obtain the following eigenvalue equation:

$$([A] - n_{\text{eff}}^2[B])\{h\} = 0 \tag{3.23}$$

where $[A]$ and $[B]$ are the finite-element matrices and $\{h\}$ is the discretized electric field vector consisting of the edge and nodal variables. The matrices $[A]$ and $[B]$ are sparse, allowing an efficient resolution of the equation to be obtained using high-performance algebraic solvers for both real and complex problems. In order to enclose the computational domain without affecting the numerical solution, anisotropic perfectly matched layers (PMLs) are placed before the outer boundary. This formulation is able to deal with anisotropic material both in terms of dielectric permittivity and magnetic permeability, allowing anisotropic PML to be directly implemented.

3.7.1 Perfectly matched layer

The PML formulation can be deduced from Maxwell's equations by introducing a complex-valued coordinate transformation under the additional requirement that the wave impedance should remain unaffected. From an implementation viewpoint, it is more practical to describe the PML as an anisotropic material with losses. The PML can have arbitrary thickness and is specified as being made of an artificial absorbent material. The material has anisotropic permittivity and permeability that match the permittivity and permeability of the physical medium outside the PML in such a way that there are no reflections [25, 26]. Figure 3.13 is a schematic of the transverse cross section of a PCF surrounded by PML regions at the edges of the computational window, where x and y are the transverse directions and z is the propagation direction. PML regions I and II face toward the x and y directions, respectively, region III corresponds to the four corners, and W is the width of the PML. If the analysis region Ω is surrounded by anisotropic PML, the permittivity and permeability tensors in the PML region can be expressed as

$$[\varepsilon]_{\text{PML}} = \varepsilon_0 n^2[L], \quad [\mu]_{\text{PML}} = \mu_0[L], \tag{3.24}$$

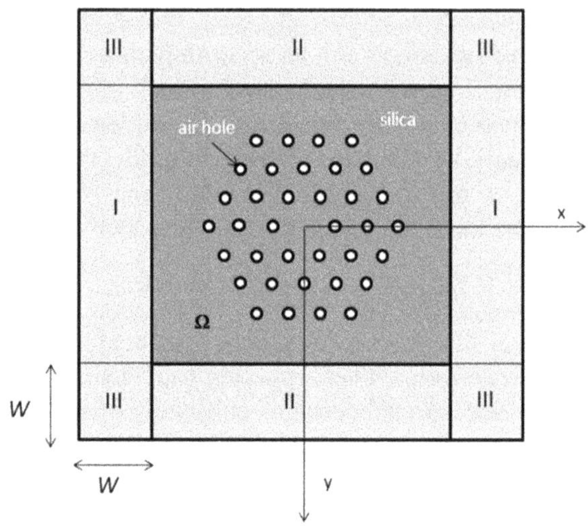

Figure 3.13. Schematic diagram of a PCF surrounded by a PML.

where ε_0 and μ_0 are the permittivity and permeability of free space, and 'n' is the refractive index with

$$[L] = \begin{bmatrix} \frac{s_y s_z}{s_x} & 0 & 0 \\ 0 & \frac{s_z s_x}{s_y} & 0 \\ 0 & 0 & \frac{s_x s_y}{s_z} \end{bmatrix}. \tag{3.25}$$

Here, s_x, s_y, and s_z are the parameters associated with the PML boundary conditions imposed at the edges of the computational window. The attenuation of the field **E** in the PML regions can be controlled by choosing the following appropriate values [25]:

$$S = 1 - j\frac{3\lambda\rho_d^2}{4\pi n W^3}\ln\left(\frac{1}{R_t}\right), \tag{3.26}$$

where ρ_d is the distance inside the PML measured from the interface between the PML and the edge of the computational window and R_t is a theoretical reflection coefficient at the interface between the PML and the edge of the computational window. The PML parameters s_x and s_y are given in table 3.1, and s_z is unity if wave propagation is assumed to take place in the z direction. In order to reduce the computational demand, structural symmetries can be exploited for the numerical simulations. This makes the modeling of realistic structures possible with high accuracy and short time delays. Recently, a number of commercial packages have appeared that can solve Maxwell's equations using a finite element combined with

Table 3.1. Parameters s_x and s_y.

PML parameter	Region I	Region II	Region III
s_x	S	1	S
s_y	1	S	S

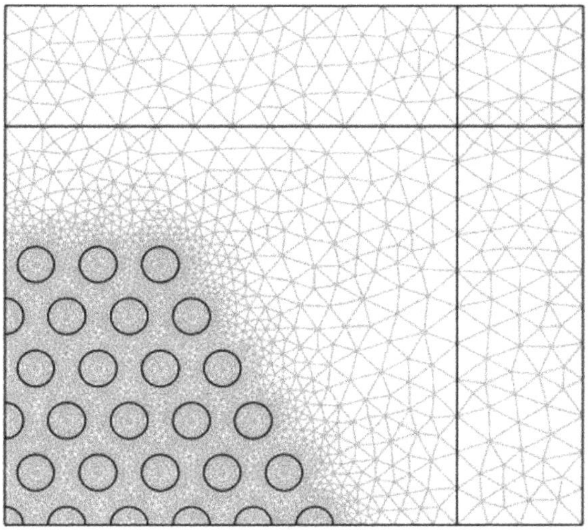

Figure 3.14. One quarter of a PCF structure with the generated mesh.

fast matrix eigenvalue solvers. For example, figure 3.14 shows one quarter of the fiber cross section, which is divided into triangular elements [27].

3.7.2 Photonic crystal fiber parameters

Using the FEM, the effective refractive index of the PCF as a function of λ can be obtained for different PCF designs, namely, S_1 ($d/\Lambda = 0.4$; $\Lambda = 3\ \mu$m), S_2 ($d/\Lambda = 0.6$; $\Lambda = 3\ \mu$m), and S_3 ($d/\Lambda = 0.6$; $\Lambda = 1\ \mu$m). Using the dispersion relation, the GVD can be calculated as a function of λ for S_1, S_2, and S_3 using equation (3.11), as shown in figure 3.15. As the air filling factor changes as a result of changing the pitch and air-hole size, the effective cladding index n_{FSM} and hence n_{eff} are also altered by different PCF designs. This wavelength dependence leads to strong variation in the dispersion parameters, an effect illustrated in figure 3.15. For silica-core PCFs with different structures, zero dispersion can be obtained for wavelengths ranging all the way from 400 nm to 1600 nm, thus providing anomalous dispersion at any wavelength, which is not possible using ordinary conventional fiber. Such PCFs are utilized for nonlinear applications such as soliton generation and propagation, soliton lasers, supercontinuum generation (SCG), and ultrashort pulse compression in the visible regime. Figure 3.15 also depicts the calculated wavelength dependence of the nonlinearity for PCF structures S_1, S_2, and S_3.

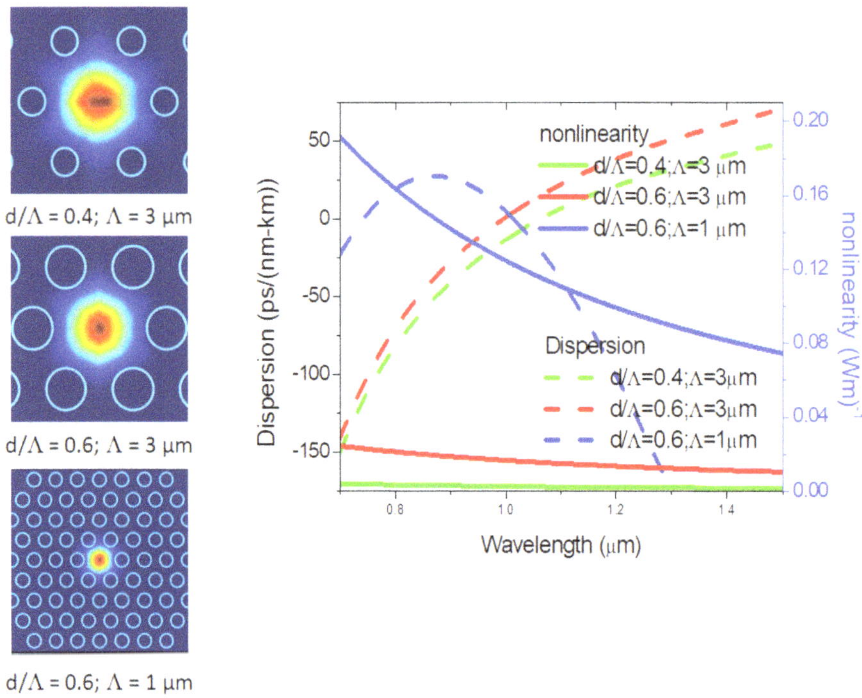

Figure 3.15. Calculation of the dispersion and nonlinear parameter of PCF using FEM.

References

[1] Agrawal G P 2006 *Nonlinear Fiber Optics* 4 (Cambridge, MA: Academic Press)

[2] Raja R V J and Porsezian K 2007 A fully vectorial effective index method to analyse the propagation properties of microstructured fiber *Photon. Nanostruct. Fundam. Appl.* **5** 171–7

[3] Knight J C, Birks T A, Russell P S J and Atkin D M 1996 All-silica single-mode optical fiber with photonic crystal cladding *Opt. Lett.* **21** 1547–9

[4] Varshney S K, Singh M P and Sinha R K 2003 Propagation characteristics of photonic crystal fibers *J. Opt. Commun.* **24** 192–8

[5] Ghosh R, Kumar A and Meunier J-P 1999 Waveguiding properties of holey fibres and effective-V model *Electron. Lett.* **35** 1873–5

[6] Fogli F, Saccomandi L, Bassi P, Bellanca G and Trillo S 2002 Full vectorial BPM modeling of index-guiding photonic crystal fibers and couplers *Opt. Express* **10** 54–9

[7] He Y Z and Shi F G 2003 Finite-difference imaginary-distance beam propagation method for modeling of the fundamental mode of photonic crystal fibers *Opt. Commun.* **225** 151–6

[8] Qiu M 2001 Analysis of guided modes in photonic crystal fibers using the finite-difference time-domain method *Microw. Opt. Technol. Lett.* **30** 327–30

[9] Zhu Z and Brown T G 2002 Full-vectorial finite-difference analysis of microstructured optical fibers *Opt. Express* **10** 853–64

[10] Guenneau S, Nicolet A, Zolla F and Lasquellec S 2002 Modeling of photonic crystal optical fibers with finite elements *IEEE Trans. Magn.* **38** 1261–4

[11] Koshiba M 2002 Full-vector analysis of photonic crystal fibers using the finite element method *IEICE Trans. Electron.* **85** 881–8 https://search.ieice.org/bin/summary.php?id=e85-c_4_881&category=C&year=2002&lang=E&abst=
[12] Zhi W, Guobin R, Shuqin L and Shuisheng J 2003 Supercell lattice method for photonic crystal fibers *Opt. Express* **11** 980–91
[13] Arriaga J, Knight J C and Russell P S J 2003 Modelling photonic crystal fibres *Physica* E **17** 440–2
[14] Fini J M 2003 Analysis of microstructure optical fibers by radial scattering decomposition *Opt. Lett.* **28** 992–4
[15] Kuhlmey B T, White T P, Renversez G, Maystre D, Botten L C, De Sterke C M and McPhedran R C 2002 Multipole method for microstructured optical fibers. II. Implementation and results *J. Opt. Soc. Am.* B **19** 2331–40
[16] Li Y-f, Wang C-y and Hu M-l 2004 A fully vectorial effective index method for photonic crystal fibers: application to dispersion calculation *Opt. Commun.* **238** 29–33
[17] Yariv A and Yeh P 2007 *Photonics: Optical Electronics in Modern Communications* (Oxford: Oxford University Press)
[18] Park K N and Lee K S 2005 Improved effective-index method for analysis of photonic crystal fibers *Opt. Lett.* **30** 958–60
[19] Bréchet F, Marcou J, Pagnoux D and Roy P J O F T 2000 Complete analysis of the characteristics of propagation into photonic crystal fibers, by the finite element method *Opt. Fiber Technol.* **6** 181–91
[20] Husakou A V and Herrmann J 2001 Supercontinuum generation of higher-order solitons by fission in photonic crystal fibers *Phys. Rev. Lett.* **87** 203901
[21] Lidiya A E, Raja R V J, Ngo Q M and Vigneswaran D *et al* 2019 Detecting hemoglobin content blood glucose using surface plasmon resonance in d-shaped photonic crystal fiber *Opt. Fiber Technol.* **50** 132–8
[22] Broeng J, Mogilevstev D, Barkou S E and Bjarklev A 1999 Photonic crystal fibers: a new class of optical waveguides *Opt. Fiber Technol.* **5** 305–30
[23] Baggett J C, Monro T M, Furusawa K and Richardson D J 2001 Comparative study of large-mode holey and conventional fibers *Opt. Lett.* **26** 1045–7
[24] Ghatak A A, Ghatak A, Thyagarajan K and Thyagarajan K 1998 *An Introduction to Fiber Optics* (Cambridge: Cambridge University Press)
[25] Obayya S S A, Rahman B M A and Grattan K T V 2005 Accurate finite element modal solution of photonic crystal fibres *IEE Proc.* **152** 241–6
[26] Saitoh K and Koshiba M 2002 Full-vectorial imaginary-distance beam propagation method based on a finite element scheme: application to photonic crystal fibers *IEEE J. Quantum Electron.* **38** 927–33
[27] COMSOL Multiphysics 1998 Introduction to comsol multiphysics *COMSOL Multiphysics, Burlington, MA* (accessed February 9, 2018)

IOP Publishing

Fiber Optic Pulse Compression
Numerical techniques and applications with MATLAB®
R Vasantha Jayakantha Raja and A Esther Lidiya

Chapter 4

Soliton propagation

We begin our manipulation of light in the spatial domain, where light is captured and guided via the intelligent arrangement of dielectric material. Once trapped, light continuously interacts with the dielectric material as it propagates through it, building even minute perturbations into a dramatic response. All of the fascinating effects that originate from the very act of light–matter interaction are discussed here in terms of induced polarization. Hence, a brief description of linear and nonlinear polarization is first given. This is then followed by a short discussion of the propagation properties of light in a dielectric waveguide. We then describe the interesting coupling between linear and nonlinear effects in an optical fiber under different circumstances, and show how they lead to the spectral broadening and soliton dynamics that facilitate the results presented in later chapters.

4.1 Soliton

The major constraints in optical fiber communication are error and cross talk. These mainly occur due to optical losses and dispersion in optical fibers. In single-mode fibers, there are two types of dispersion that hinder communications applications, namely, material dispersion and waveguide dispersion. As no source of light is purely monochromatic, different wavelengths experience different refractive index values and hence propagate at different velocities, with the result that the propagating wavelengths of an optical pulse reach the end of the fiber at different timings. This type of dispersion is known as material dispersion. Waveguide dispersion occurs due to the structural design of a fiber, i.e. the size of the core and the refractive index difference between the core and the cladding. The combined effects of material dispersion and waveguide dispersion, referred to as group velocity dispersion (GVD), ultimately result in pulse temporal broadening. Dispersion make the pulse energy spread temporally, which leads to energy loss. However, when this dispersive wave propagates through a non-dissipative system with an amplitude larger than a threshold level (to induce nonlinearity in the system), there is a

doi:10.1088/978-0-7503-2686-5ch4

possibility that the dispersion and the nonlinear effect can exactly balance, leading to the formation of stable pulses called solitons. Theoretically, solitons are confined energy pulses which can travel an infinite distance without any change in shape.

A soliton was first observed by Scott Russell in 1834 [1]. While he was observing a boat moving through a shallow canal, it was suddenly stopped by a 'heap' of water which propagated over couple of kilometers without distortion. He was so impressed by this great solitary wave that he spent ten years of his life experimentally studying the properties of this singular and beautiful phenomenon. However, such wave properties were not completely understood until the works of Boussinesq in 1872 [2] and Korteweg and de Vries (K–dV) in 1895 [3]. The phenomenon was then forgotten until 1965, when Zabusky and Kruskal coined the word **soliton** to describe the particle-like properties of the solitary wave [4]. They also provided a fundamental understanding of the origin of the soliton and initiated the development of mathematical methods in order to study its behavior.

Later, in 1973, the existence of optical solitons was predicted by Hasegawa and Tappert [5] and experimentally confirmed by Mollenaur and his group in 1980 [6]. They had theoretically proven the possibility of the generation and propagation of solitons in optical fiber that results from balancing dispersion and nonlinearity. Optical solitons can be modified to be more immune to external perturbations by applying appropriate variations in the fiber dispersion profile, hence they can be effectively utilized for communications applications. An optical soliton is a short, bell-shaped laser pulse that can travel through a fiber for thousands of kilometers without substantial dispersion if the system's loss is taken into account. These solitons can only be realized in the nonlinear regime. The development of soliton envelopes in anomalous and normal dispersive fibers is referred to as the development of bright and dark solitons, respectively [7]. Bright solitons are most often used in optical fiber transmission systems. The existence of solitons has been proved in many areas of science, namely, particle physics [8], magnetic films [9], molecular biology [10], quantum mechanics [11], meteorology [12], oceanography [13], astrophysics [7], and cosmology [14]. However, the solitons that exist in optics (referred to as 'optical solitons') have piqued the scientific community's interest, because they appear to be ideal candidates for transporting information (music, video, or data) across the globe via optical fibers. In addition to temporal optical solitons, a variety of novel solitons have been discovered, including vortex solitons [15], spatial vector solitons [16], bound solitons [17], blue solitons [18], Bragg and gap solitons [19], discrete solitons [7], breather solitons [20], etc.

4.2 Nonlinear propagation in optical fiber

4.2.1 Polarization response

Light–matter interactions are generically categorized into linear and nonlinear effects, based on whether nonlinear coupling occurs between distinct light frequency components or not. Material responses that are unaffected by other frequencies are thus investigated in terms of their linear impacts. Material responses, on the other hand, are classified as nonlinear when they are affected by the existence of additional frequency components, such as when they are reliant on total field intensity. When such light propagates through a dielectric material such as a waveguide, it disrupts the bound electrons in the substrate, inducing a polarization response. In particular, an oscillating electric field, such as that caused by the passage of light, drives the bound electrons into oscillation, in analogy to a forced harmonic oscillator. This creates a time-dependent polarization represented by

$$P_L(r,\,t) = \epsilon_0 \int_{-\infty}^{t} \chi^{(1)}(t - t')E(r,\,t')dt', \tag{4.1}$$

where $E(r,\,t)$ is the electric field in the time domain, ϵ_0 is the permittivity of free space, and $\chi^{(1)}$ is the electric susceptibility, which is generally a tensor. Equation (4.1) represents a good approximation to the polarization response of a dielectric material. Later, the invention of the laser and the resulting intense coherent radiation pushed the linear model beyond its limit. Upon the incidence of laser light, the nonlinear response of the bound electrons due to the complexity of the electron levels influences the atomic potential. Accordingly, a quantum-mechanical picture gives a more exact analysis of the origin of this nonlinearity. However, when the driving electric field is far from oscillatory resonance, the nonlinear effects can be treated as a small but important perturbation, which is given by

$$\begin{aligned} P_{\mathrm{NL}}(r,\,t) = \epsilon_0 &\iint_{-\infty}^{t} \chi^{(2)}(t - t_1,\, t - t_2) \colon \times\, E(r,\,t_1)E(r,\,t_2)dt_1dt_2 \\ + \epsilon_0 &\iiint_{-\infty}^{t} \chi^{(3)}(t - t_1,\, t - t_2,\, t - t_3) \times E(r,\,t_1)E(r,\,t_2)E(r,\,t_3)dt_1dt_2dt_3, \end{aligned} \tag{4.2}$$

which represents the additional second and third orders of nonlinear polarization. The importance and characteristics of even higher-order terms are still not conclusively known. Hence, the induced polarization can be represented as:

$$\mathbf{P}(\mathbf{r},\,t) = \mathbf{P}_L(\mathbf{r},\,t) + \mathbf{P}_{\mathrm{NL}}(\mathbf{r},\,t) \tag{4.3}$$

$$\begin{aligned} \mathbf{P}(\mathbf{r},\,t) = \epsilon_0 &\int_{-\infty}^{\infty} \chi^{(1)}(t - t_1)\mathbf{E}(\mathbf{r},\,t_1)dt_1 \\ + \epsilon_0 &\int_{-\infty}^{\infty}\int_{-\infty}^{\infty} \chi^{(2)}(t - t_1,\, t - t_2)\mathbf{E}(\mathbf{r},\,t_1)\mathbf{E}(\mathbf{r},\,t_2)dt_1dt_2 \\ + \epsilon_0 &\int_{-\infty}^{\infty}\int_{-\infty}^{\infty}\int_{-\infty}^{\infty} \chi^{(3)}(t - t_1,\, t - t_2,\, t - t_3)\mathbf{E}(\mathbf{r},\,t_1)\mathbf{E}(\mathbf{r},\,t_2)\mathbf{E}(\mathbf{r},\,t_3)dt_1 \\ &dt_2dt_3 + \cdots, \end{aligned} \tag{4.4}$$

where $\chi^{(j)}$ is the jth-order susceptibility and $|P_{\mathrm{NL}}| < <|P_L|$, even at high power. If the medium's response is instantaneous compared to the pulse duration ($t_1/T_0 \ll 1$, where T_0 is the pulse width and t_1 is the nonlinear response time of the medium) then equation (4.4) may be approximated by

$$\mathbf{P}(\mathbf{r},\,t) = \epsilon_0[\chi^{(1)} \cdot \mathbf{E}(\mathbf{r},\,t) + \chi^{(2)} \colon \mathbf{E}(\mathbf{r},\,t)\mathbf{E}(\mathbf{r},\,t) + \chi^{(3)} \vdots\, \mathbf{E}(\mathbf{r},\,t)\mathbf{E}(\mathbf{r},\,t)\mathbf{E}(\mathbf{r},\,t) + \cdots]. \tag{4.5}$$

As discussed earlier, linear effects are due to $\chi^{(1)}$; $\chi^{(2)}$ is the second-order susceptibility. In addition, in dielectric materials possessing centrosymmetry, such as silica glass or a bulk gas, all the even-order susceptibilities vanish, thus the contribution of second-order susceptibility is negligible. Hence, in silica fiber, the lowest dominant nonlinearity present is due to $\chi^{(3)}$ as it is made up of SiO_2 molecules. Hence, the linear and a nonlinear response can be written as

$$\mathbf{P}_L(\mathbf{r}, t) = \varepsilon_0\chi^{(1)}\mathbf{E}(\mathbf{r}, t) \tag{4.6}$$

$$\mathbf{P}_{\mathrm{NL}}(\mathbf{r}, t) = \varepsilon_0\chi^{(3)}\mathbf{E}(\mathbf{r}, t)\mathbf{E}(\mathbf{r}, t)\mathbf{E}(\mathbf{r}, t). \tag{4.7}$$

For the silica material, the relation represented in equation (4.2) can be presented more simply when written in the frequency domain. For example, the linear and nonlinear polarizations can be represented by

$$\tilde{\mathbf{P}}_L(r, \omega) = \epsilon_0\chi^{(1)}(\omega)\tilde{\mathbf{E}}(r, \omega) \tag{4.8}$$

$$\mathbf{P}_{\mathrm{NL}}(\mathbf{r}, \omega) = \varepsilon_0\chi^{(3)}(\omega)\mathbf{E}(\mathbf{r}, \omega)\mathbf{E}(\mathbf{r}, \omega)\mathbf{E}(\mathbf{r}, \omega), \tag{4.9}$$

where $\tilde{P}_L(r, \omega)$, $\tilde{P}_{\mathrm{NL}}(r, \omega)$, $\tilde{E}(r, \omega)$, and $\chi^{(1)}(\omega)$ are the Fourier transforms of P_L(r,t), P_{NL}(r,t), E(r,t), and $\chi^{(1)}(t)$, respectively. The relative permittivity of the medium is represented by $\epsilon(r) = 1 + \chi^{(1)}(\omega)$. This allows us to write the electric displacement in the form

$$\tilde{\mathbf{D}}(r, \omega) = \epsilon_0\epsilon(\omega)\tilde{\mathbf{E}}(r, \omega) + \tilde{\mathbf{P}}_{\mathrm{NL}}(r, \omega). \tag{4.10}$$

4.2.2 Nonlinear Schrödinger equation

Electromagnetic wave propagation in any medium is governed by Maxwell's equations [7, 21]. From Maxwell's equations, by considering the linear and nonlinear effects, the pulse envelope equation can be derived in the form of the nonlinear Schrödinger equation (NLSE):

$$\nabla \times \mathbf{B} = \mathbf{J} + \frac{\partial \mathbf{D}}{\partial t} \tag{4.11}$$

$$\nabla \times \mathbf{E} = -\frac{\partial \mathbf{B}}{\partial t} \tag{4.12}$$

$$\nabla . \mathbf{B} = 0 \tag{4.13}$$

$$\nabla . \mathbf{D} = \rho, \tag{4.14}$$

where **H** is the magnetic field vector, **E** is the electric field vector, and **B** and **D** are the magnetic and electric flux densities, respectively. The current density vector is **J** and the charge density is ρ. In the optical fiber, **J**, ρ and **M** are zero. Therefore, the electric and magnetic flux density vectors are defined as

$$\mathbf{D} = \epsilon_0\mathbf{E} + \mathbf{P} \tag{4.15}$$

$$\mathbf{B} = \mu_0\mathbf{H}, \tag{4.16}$$

where μ_0 is the permeability of free space, ϵ_0 is the permittivity of free space and **P** and **M** are the induced electric and magnetic polarizations. By taking the curl of equation (4.12) and using equation (4.11), (4.15), and (4.16), one can obtain

$$\nabla \times \nabla \times \mathbf{E}(\mathbf{r}, t) + \frac{1}{c^2}\frac{\partial^2 \mathbf{E}(\mathbf{r}, t)}{\partial t^2} + \mu_0 \frac{\partial^2 \mathbf{P}(\mathbf{r}, t)}{\partial t^2} = 0, \quad (4.17)$$

where $c\ (= 1/\sqrt{\mu_0\,\epsilon_0})$ is the light velocity in vacuum. This equation reduces to

$$\nabla^2 \mathbf{E} - \frac{1}{c^2}\frac{\partial^2 \mathbf{E}}{\partial t^2} = \mu_0 \frac{\partial^2 \mathbf{P}}{\partial t^2}. \quad (4.18)$$

Utilizing equation (4.5), one can determine the polarization response of the dielectric material.

4.2.3 Deriving the nonlinear Schrödinger equation

In order to derive the basic propagation equation, it is necessary to make several simplifying assumptions before solving equation (4.18)

- $\mathbf{P}_{\mathrm{NL}}$ is treated as a small perturbation to $\mathbf{P}_L$. This is justified because nonlinear effects are weak in silica fibers.
- The optical field is assumed to maintain its polarization along the fiber length so that a scalar approach is valid.
- The optical field is assumed to be quasi-monochromatic, i.e. the pulse spectrum, centered at ω_0, is assumed to have a spectral width of $\Delta\omega$ such that $\Delta\omega/\omega_0 \ll 1$.

According to the slowly varying envelope approximation, it is useful to separate the rapidly varying part of the electric field by writing it in the form of

$$\mathbf{E}(\mathbf{r}, t) = \frac{1}{2}\hat{x}[U(\mathbf{r}, t)e^{-i\omega_0 t} + U^*(\mathbf{r}, t)e^{i\omega_0 t}], \quad (4.19)$$

where $\hat{x}$ is the polarization unit vector of the light (assumed to be linearly polarized along the x-axis), $U(\mathbf{r}, t)$ is a slowly varying function of time (relative to the optical period) and U^* denotes the complex conjugate of U. Equation (4.19) is substituted into equation (4.6) and (4.9) and a similar form to that used in equation (4.19) is used for the polarization vector :

$$\mathbf{P}_L(\mathbf{r}, t) = \frac{1}{2}\hat{x}[P_L(\mathbf{r}, t)e^{-i\omega_0 t} + P_L^*(\mathbf{r}, t)e^{i\omega_0 t}] \quad (4.20)$$

$$\mathbf{P}_{\mathrm{NL}}(\mathbf{r}, t) = \frac{1}{2}\hat{x}[P_{\mathrm{NL}}(\mathbf{r}, t)e^{-i\omega_0 t} + P_{\mathrm{NL}}^*(\mathbf{r}, t)e^{i\omega_0 t}]. \quad (4.21)$$

A Fourier transform is applied to equation (4.18), and equation (4.20) is substituted into equation (4.21), which expresses $P_L(\mathbf{r}, t)$ and $P_{\mathrm{NL}}(\mathbf{r}, t)$ with their relation to $U(\mathbf{r}, t)$. The obtained wave equation has the form of

$$\nabla^2 \tilde{U} + \varepsilon(\omega) k_0^2 \tilde{U} = 0, \tag{4.22}$$

where $\tilde{U}$ denotes the Fourier transform of $U(\mathbf{r}, t)$, $k_0 = \omega_0/c$ and

$$\varepsilon(\omega) = 1 + \tilde{\chi}_{xx}^{(1)}(\omega) + \varepsilon_{\mathrm{NL}}, \tag{4.23}$$

whose nonlinear part $\varepsilon_{\mathrm{NL}}$ can be written as

$$\varepsilon_{\mathrm{NL}} = \frac{3}{4}\tilde{\chi}_{xxxx}^{(3)} |U(\mathbf{r}, t)|. \tag{4.24}$$

Equation (4.22) is known as the Helmholtz equation, and it can be solved by using the method of separation of variables:

$$\tilde{U}(\mathbf{r}, \omega - \omega_0) = F(x, y)\tilde{U}(z, \omega - \omega_0)e^{i\beta_0 z} \tag{4.25}$$

where $\tilde{U}(\mathbf{r}, \omega - \omega_0)$ is a slowly varying function of z, β_0 is the wave number, and $F(x, y)$ is a function which corresponds to the transverse electric modes in the (x, y) plane. If the z-axis is identical to the propagation direction, equation (4.22) leads to the following two equations for $F(x, y)$ and $\tilde{U}$:

$$\frac{\partial^2 F}{\partial x^2} + \frac{\partial^2 F}{\partial y^2} + \left(\varepsilon_0(\omega)k_0^2 - \bar{\beta}^2\right)F = 0, \tag{4.26}$$

$$\frac{\partial^2 \tilde{U}}{\partial z^2} + 2i\beta_0 \frac{\partial \tilde{U}}{\partial z} + \left(\bar{\beta}^2 - \beta_0^2\right)\tilde{U} = 0. \tag{4.27}$$

In obtaining equation (4.27), the second derivative can be neglected since $\tilde{U}(z, \omega)$ is assumed to be a slowly varying function of z. The wave number $\bar{\beta}$ is determined by solving the eigenvalue equation (4.26). The eigenvalue $\bar{\beta}$ can be written in the form of

$$\bar{\beta}(\omega) = \beta(\omega) + \Delta\beta, \tag{4.28}$$

where $\Delta\beta$ is a perturbation term and $\beta(\omega)$ is the frequency-dependent mode-propagation constant. Thus, from equation (4.27) we obtain

$$\frac{\partial \tilde{U}}{\partial z} - \frac{i}{2}\left\{[\beta(\omega)^2 + 2\beta(\omega)\Delta\beta]\frac{1}{\beta_0} - \beta_0\right\}\tilde{U} = 0. \tag{4.29}$$

It is useful to expand $\beta(\omega)$ in a Taylor series around the carrier frequency ω_0, as follows:

$$\beta - \beta_0 = \left.\frac{\partial \beta}{\partial \omega}\right|_{\omega_0} (\omega - \omega_0) + \frac{1}{2}\left.\frac{\partial^2 \beta}{\partial \omega^2}\right|_{\omega_0} (\omega - \omega_0)^2 + \frac{1}{6}\left.\frac{\partial^3 \beta}{\partial \omega^3}\right|_{\omega_0} (\omega - \omega_0)^3 + \cdots. \tag{4.30}$$

Neglecting the terms that are higher than second order, we take the inverse Fourier transform using

$$U(z, t) = \frac{1}{2\pi} \int_{-\infty}^{\infty} \tilde{U}(z, \omega - \omega_0) \exp\left[-i(\omega - \omega_0)t\right] d\omega. \qquad (4.31)$$

The resulting equation for $U(z, t)$ becomes

$$\frac{\partial U}{\partial z} + \beta_1 \frac{\partial U}{\partial t} + i\frac{\beta_2}{2}\frac{\partial^2 U}{\partial t^2} - i\Delta\beta U = 0. \qquad (4.32)$$

The term with $\Delta\beta$ includes the effects of fiber loss and nonlinearity. It can be evaluated from equation (4.26) using first-order perturbation theory:

$$\Delta\beta = -\frac{\alpha}{2} + i\gamma|U|^2, \qquad (4.33)$$

where γ is the nonlinear coefficient defined by

$$\gamma = \frac{n_2\omega_0}{cA_{\text{eff}}}, \qquad (4.34)$$

n_2 is the nonlinear refractive index, and A_{eff} is the effective core area, which is given in the form of

$$A_{\text{eff}} = \frac{\left(\int\int_{-\infty}^{\infty} |F(x, y)|^2 dxdy\right)^2}{\int\int_{-\infty}^{\infty} |F(x, y)|^4 dxdy}, \qquad (4.35)$$

where $F(x, y)$ is the transverse-mode field distribution. By substituting equation (4.33) into equation (4.32) and making a variable transformation with $T = t - \beta_1 z$, one can obtain the NLSE as [7]

$$\frac{\partial U(z, T)}{\partial z} = -\frac{\alpha}{2}U - i\frac{\beta_2}{2}\frac{\partial^2 U}{\partial T^2} + i\gamma|U|^2 U. \qquad (4.36)$$

Attenuation is described by the first term on the right-hand side of equation (4.36), GVD corresponds to the second term, and nonlinearity or self-phase modulation (SPM) is described by the third term with the intensity dependence.

4.2.4 Higher-order nonlinear effects

Equation (4.36) does not include inelastic scattering, such as Raman or Brillouin scattering, which becomes important above a threshold of pulse peak intensity and below a certain pulse duration. In particular, the Raman effect is usually important below the 1 ps timescale if the Raman threshold is reached. Below 1 ps, the spectral width can be broad enough that the Raman gain transfers energy from the low-frequency components to the higher-frequency components. This results in a self-frequency shift of the pulse, whose physical origin is the delayed nature of the Raman response. In this approximation, the nonlinear response of the medium is comparable with the pulse width. Thus, the intensity-dependent nonlinear effects can be included by assuming the following form for the third-order susceptibility

$$\chi^{(3)}(t - t_1, t - t_2, t - t_3) = \chi^{(3)} R(t - t_1)\delta(t - t_2)\delta(t - t_3), \tag{4.37}$$

where $R(t)$ is the nonlinear response function normalized in the same way as the delta function. Higher-order dispersion terms can easily be added, including higher-order Taylor coefficients from the expansion of $\beta(\omega)$ during the derivation process of (4.36) at the step represented by equation (4.29). Substituting equation (4.37) into equation (4.4) and performing a similar derivation process to that used for equation (4.36) yields

$$\begin{aligned}\frac{\partial U(z, t)}{\partial z} = -\frac{\alpha}{2}U - \sum_{m=1}^{M}\beta_m \frac{i^{m-1}}{m!}\frac{\partial^m U}{\partial t^m} + i\gamma \\ \left(1 + \frac{i}{\omega_0}\frac{\partial}{\partial t}\right)\left(U\int_{-\infty}^{\infty} R(t')|U(z, t - t')|^2 dt'\right).\end{aligned} \tag{4.38}$$

The response function $R(t)$ includes the electronic (instantaneous) and vibrational (delayed) Raman responses:

$$R(t) = (1 - f_R)\delta(t) + f_R h_R(t), \tag{4.39}$$

where f_R is the fractional contribution of the delayed Raman response to the nonlinear polarization and $h_R(t)$ is the Raman response function. Using the assumption that $\Delta t_1 >> 10$ fs, equation (4.38) can be simplified to the following expression [7]:

$$\frac{\partial U}{\partial z} + \frac{\alpha}{2}U + \sum_{n=2}^{4}\beta_n \frac{i^{n-1}}{n!}\frac{\partial^n U}{\partial T^n} = i\gamma\left(|U|^2 U + \frac{i}{\omega_0}\frac{\partial(|U|^2 U)}{\partial T} - T_R U\frac{\partial |U|^2}{\partial T}\right), \tag{4.40}$$

where

$$T_R \approx f_R \int_0^{\infty} t^* h_R(t) dt, \tag{4.41}$$

f_R represents the fractional contribution of the delayed Raman response, and h_R is the Raman response function, which can be approximately represented as

$$h_R = \frac{\tau_1^2 + \tau_2^2}{\tau_1 \tau_2^2}\sin\left(\frac{t}{\tau_1}\right)\exp\left(-\frac{t}{\tau_2}\right), \tag{4.42}$$

where τ_1 and τ_2 are adjusting parameters with typical values for silica of 12.2 fs and 32 fs, respectively. Using this form of the Raman response function, the integration of equation (4.41) can be performed analytically [7]:

$$T_R \approx f_R \frac{2\tau_1^2 \tau_2}{\tau_1^2 + \tau_2^2}. \tag{4.43}$$

This can be used to approximate the Raman scattering effect in the last term of equation (4.40).

4.3 Split-step Fourier method

The NLSE as given in equation (4.38) cannot be solved analytically for the general case of arbitrarily shaped pulses launched into a fiber. However, powerful numerical procedures to solve it have been developed over the years. Two main types of numerical scheme are generally applied to solve the NLS equation: finite-difference methods, such as the Crank-Nicolson [22] and Ablowitz and Ladik [23, 24] schemes, and function approximation methods, such as the split-step Fourier method (SSFM). Among them, the SSFM has shown to be the most reliable technique since the study of Hasegawa and Tappert in 1973 [5]. Later, the SSFM was the most extensively utilized technique because of its ease of implementation and speed when compared to other approaches, particularly time-domain finite-difference methods [7]. The SSFM is the methodology of choice for solving the NLS equation. First, the mathematical terms representing the effects of dispersion and nonlinearity are decoupled from the NLS equation as follows:

$$\hat{D} = -\frac{\alpha}{2}U - \sum_{n=2}^{4}\beta_n \frac{i^{n-1}}{n!}\frac{\partial^n U}{\partial T^n} \tag{4.44}$$

$$\hat{N} = i\gamma\left(|U|^2 U + \frac{i}{\omega_0}\frac{\partial(|U|^2 U)}{\partial T} - T_R U\frac{\partial|U|^2}{\partial T}\right). \tag{4.45}$$

Here, the operators $\hat{D}$ and $\hat{N}$ can be considered separately from each other over short fiber distances Δz. where $U(z, T)$ is the complex field envelope at step z and time T. The NLSE then can be written in the operator form as

$$\frac{\partial U}{\partial z} = (\hat{D} + \hat{N})U. \tag{4.46}$$

The solution to this equation is based on splitting each step Δz of the numerical integration so that only the nonlinear terms in the equations are considered in the first substep, and only the GVD and loss terms are addressed at the second substep. At the latter stage, the corresponding linear equation(s) are solved by means of a Fourier transform.

In general, dispersion and nonlinearity acting together along the length of the fiber. In order to determine a more approximate solution, SSFM assumes that the dispersive and nonlinear effects act independently while propagating through the optical fiber over a small distance h. More specifically, propagation from z to $z + h$ is carried out in two steps. In the first step, nonlinearity acts alone, and $\hat{D} = 0$ in equation (4.38) as shown in figure 4.1. In the second step, dispersion acts alone, and $\hat{N} = 0$ in equation (4.38). Mathematically,

$$U(z + h, T) \approx \exp(h\hat{D})\exp(h\hat{N})U(z, T). \tag{4.47}$$

The exponential operator $\exp(h\hat{D})$ can be evaluated in the Fourier domain using the prescription

$$\exp(h\hat{D})B(z, T) = F_T^{-1} \exp\left(h\hat{D}(-i\omega)\right)F_T B(z, T), \tag{4.48}$$

where F_T denotes the Fourier transform operation, $\hat{D}(-i\omega)$ is obtained from equation (4.44) by replacing the operator $\partial/\partial T$ with $-i\omega$, and ω is the frequency in the Fourier domain. As $\hat{D}(i\omega)$ is just a number in the Fourier space, the evaluation of equation (4.48) is straightforward. The linear operator $\hat{D}$ is most efficiently solved in the spectral domain, while the nonlinear operator $\hat{N}$ is more favorably solved in the time domain. Assuming a discrete signal description in the time and frequency domain, the fast Fourier transform (FFT) is used for transformations from one domain to another. The use of the FFT algorithm makes the numerical evaluation of equation (4.48) relatively fast (by up to two orders of magnitude) compared with most finite-difference schemes. The accuracy of the SSFM can be improved by adopting a different procedure to propagate the optical pulse over one segment from z to $z + h$. In this procedure, equation (4.47) is replaced by

$$U(z + h, T) \approx \exp\left(\frac{h}{2}\hat{D}\right)\exp\left(\int_z^{z+h} \hat{N}(z')dz'\right)\exp\left(\frac{h}{2}\hat{D}\right)U(z, T). \tag{4.49}$$

The efficiency of the split-step method depends on both the time (or frequency) domain resolution and on the distribution of step sizes along the fiber. If Δz, the so-called split-step size, becomes too large, the condition for the separate calculations of $\hat{D}$ and $\hat{N}$ breaks, and the algorithm delivers wrong results. Thus, careful determination of the optimum split-step size is important in order to use the minimal computational effort for a given accuracy. Typically, the step size Δz should be a small fraction, thereby requiring >1000 steps/(shortest linear or nonlinear length) along the fiber length. As the speed of the FFT is proportional to N_t log2 N_t, where N_t is the number of signal samples in the time or frequency domain, careful determination of the simulation bandwidth and the time window is important to minimize the computational effort. For temporal and spectral computation windows spanning T_{span} and F_{span}, respectively, the sampling theorem imposes the condition $T_{\text{span}}F_{\text{span}}=N_p$, where N_P is the number of discretization points. For the simulation of

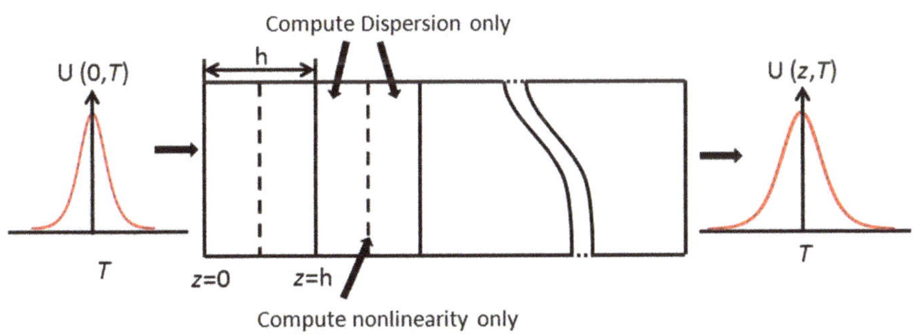

Figure 4.1. Schematic illustration of the symmetrized SSFM used for numerical simulations. The fiber length is divided into a large number of segments of width h. Within a segment, the effect of nonlinearity is included at the midplane shown by a dashed line.

femtosecond pulses, $F_{span} \sim 1000$ THz and $T_{span} \sim 20$ ps, so that a typical requirement for a very large number of points $Np \geqslant 2^{15}$ can be met.

4.4 Nonlinear propagation in optical fiber

4.4.1 Linear and nonlinear effects of fiber

(i) Dispersion

When an optical pulse that has a spectrum of Fourier frequency components propagates through a waveguide, it suffers pulse broadening due to dispersion. In general, refractive index is a measure of the velocity of wave propagation in a dielectric. Since the index of refraction of any optical system is a function of frequency, the various Fourier components of the pulse experience different indices of refraction in a dielectric medium, and hence travel with different velocities, called the group velocity. As a result of this, the optical pulse spreads in the time domain during the course of propagation. This phenomenon is called GVD or chromatic dispersion. During pulse propagation, pulse broadening is mainly due to the GVD parameter β_2. Hence, for pulse propagation in a dielectric medium, the effect of β_2 is normally considered alone. This GVD parameter varies with respect to the wavelength, λ. For one wavelength ($\lambda = 1.27$ μm) in silica, β_2 vanishes; this wavelength is called the zero dispersion wavelength (ZDW) (λ_D). This can be varied by the use of some dopants, such as GeO_2 or P_2O_5. By taking advantage of dispersion-shifted fiber, the ZDW can be shifted to the vicinity of 1.55 μm, where the fiber loss is minimal. Linear optical communication using very-low-intensity light pulses is usually based on this ZDW, since pulse broadening is very low around this wavelength. The third derivative $\beta_3\left(=\frac{\partial^3\beta}{\partial\omega^3}\right)$ is called third-order dispersion (TOD) or the higher-order dispersion parameter. When compared with β_2, the contribution of β_3 is negligible. However, for ultrashort pulses, the effect of β_3 is considerable around the ZDW.

There are two kinds of pulse broadening regime that exist on either side of λ_D. For wavelengths below λ_D ($\lambda < \lambda_D$), the red-shifted components of a pulse travel faster than the blue-shifted components. This is called the normal or positive dispersion regime. In contrast, for wavelengths above λ_D, the dispersion is known as anomalous or negative dispersion, in which the red-shifted components travel slower than the blue-shifted components. In order to account for the dispersion due to GVD, a useful parameter, called the dispersion length, is defined in the following way:

$$L_D = \frac{T_0^2}{|\beta_2|}, \tag{4.50}$$

where T_0 is the Gaussian pulse width. GVD broadens the pulse symmetrically in the cases of Gaussian pulses and hyperbolic secant pulses. Figure 4.2(a) gives an overview of the the pulse broadening of a Gaussian-shaped pulse in a fiber due to GVD. The broadening of the pulse due to GVD adds a chirp to the pulse. The chirp due to the effect of GVD is either positive or negative, depending on whether the fiber is normally

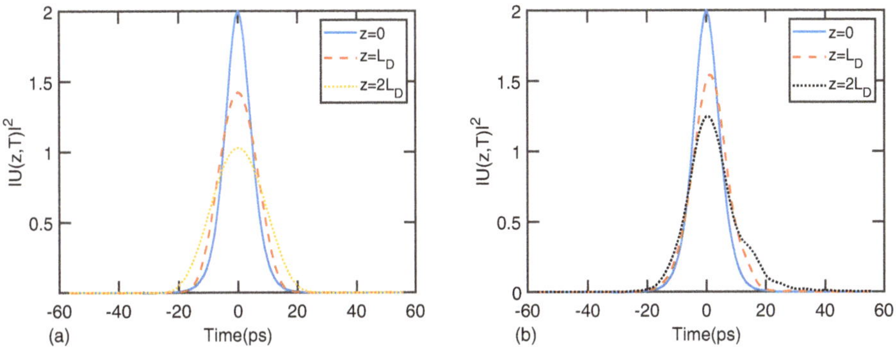

Figure 4.2. (a) Broadening of a Gaussian pulse 10 ps wide at $z = 0$, $z = L_D$, and $z = 2L_D$ for $\beta_2 = -0.0375\ \text{ps}^2\ \text{m}^{-1}$ and $\beta_3 = 0$. (b) The impact of third-order dispersion in the broadening of a Gaussian pulse at $z = 0$, $z = L_D$, and $z = 2L_D$ for $\beta_2 = -0.0375\ \text{ps}^2\ \text{m}^{-1}$ and $\beta_3 = -0.266\,39\ \text{ps}^3\ \text{m}$.

dispersive or anomalously dispersive. Thus, in both types of fiber, a propagating pulse (with zero chirp) is broadened and becomes chirped due to fiber dispersion.

(ii) Self-phase modulation

From the discussion of linear effects, it is clear that the dielectric properties depend only on the frequency of the optical pulse. However, for very intense electromagnetic fields (typically for a laser source), the behavior of a dielectric is eventually nonlinear [7, 21]. In other words, the dielectric properties also depend on the intensity of the light. Hence, the index of refraction varies with respect to intensity. This dependence is described using the nonlinear index n_2 by the relation $n = n_0 + n_2 I$, where $I = |E|^2$. This means that the phase velocity $v_p = c/n$ becomes intensity dependent, and therefore the pulse's phase is modulated by its own intensity profile. This is referred to as SPM. Hence, SPM can be defined as the phase change of an optical pulse due to self-induced change in the nonlinear refractive index. To illustrate the effects of SPM, all other terms are neglected in equation (4.40), which yields:

$$\frac{\partial A}{\partial z} = \frac{i}{L_{\text{NL}}} \exp(-2\alpha z)|A|^2 A. \tag{4.51}$$

Here, A is the normalized amplitude, which is defined as

$$A(z, T) = \sqrt{P_0}\exp(\alpha z/2)U(z, T), \tag{4.52}$$

where P_0 is the input power. The nonlinear length $L_{\text{NL}} = \frac{1}{\gamma P_0}$ gives the distance at which nonlinear effects are important. The solution of equation (4.51) is

$$A(z, T) = A(0, T)\exp\left[i\phi_{\text{NL}}(z, T)\right], \tag{4.53}$$

with

$$\phi_{\text{NL}}(z, T) = |A(0, T)|^2 \frac{z_{\text{eff}}}{L_{\text{NL}}}. \tag{4.54}$$

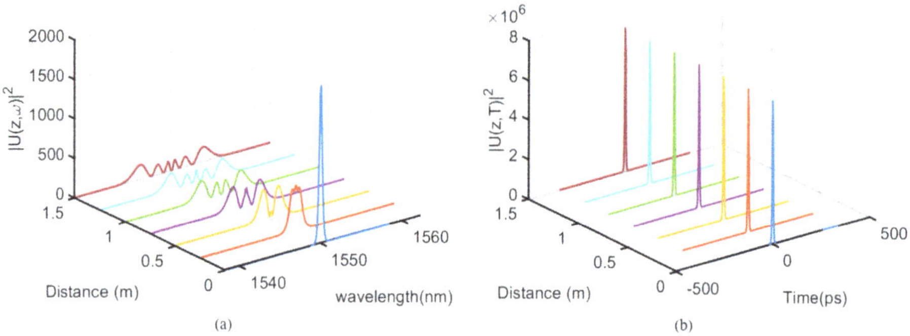

Figure 4.3. Influence of SPM in (a) the spectral evolution of an unchirped Gaussian pulse, and (b) the temporal evolution of an unchirped Gaussian pulse.

Here,

$$z_{\mathrm{eff}} = \frac{1}{\alpha}\Big[1 - \exp(-2\alpha z)\Big] \tag{4.55}$$

and it is smaller than z, thus indicating that loss limits the impact of SPM. The time-dependent frequency shift induced by the time-dependent phase ϕ_{NL} is given by

$$\delta\omega(T) = \frac{\partial|A(0,\ T)|^2}{\partial T}\frac{z_{\mathrm{eff}}}{L_{\mathrm{NL}}}. \tag{4.56}$$

The impact of SPM also induces chirp, which increases in magnitude during propagation. Over a large central region, the chirp is almost linear and positive. It can also be seen that the temporal pulse shape $|A|^2$ is unchanged during propagation. To illustrate the change in pulse spectrum due to SPM alone, equation (4.51) has been solved for a Gaussian input pulse. Figure 4.3(a) shows the spectral evolution of SPM for an unchirped Gaussian input pulse. Due to the nonlinear temporal dependence of the chirp, the nonlinear phase shift translates into broadening of the optical spectrum, generating new frequencies as the pulse travels inside the fiber. In addition, it can be observed that the frequency chirp's slope increases with distance. However, the temporal shape of the pulse remains unaffected.

(iii) Raman effect

Another nonlinear phenomenon present in optical fiber due to the propagation of very intense light is the Raman effect. When a high-intensity light wave with a carrier frequency of ω_0 is incident in a fiber that has a resonance level of ω_R, the incident light results in a downshift of the carrier frequency, generating an entirely new frequency $\omega_0 - \omega_R$, known as the mode frequency. This process is known as the Raman effect [7]. The principle behind this effect is that a lower-wavelength pump-laser pulse travels down an optical fiber along with the signal, scatters off atoms in the fiber, loses some energy to the atoms, and then continues its journey with the same wavelength as that of the signal. The signal therefore gains additional photons and is amplified. These new photons can also be joined with many more from the pump, which continue to be scattered as they travel down the fiber in a cascading process. When the frequency beating between the incident and scattered waves collectively enhances the optical photons, the scattering process is stimulated and the amplitude of the scattered wave grows exponentially in the direction of propagation. This phenomenon is known as stimulated Raman scattering (SRS) [25]. Raman gain, which results from SRS, depends on the frequency separation between the pump and the Stokes modes. Using the Raman effect, an extremely broad gain spectrum in the frequency range of about 20 THz has been obtained [26]. Raman amplification [7] aims to boost the power of signals. It is usually implemented as distributed amplification, i.e. it happens throughout the length of an actual transmission fiber. For solitons, this effect shifts the soliton spectrum toward lower frequencies, which is called a self-frequency shift [7].

4.4.2 Soliton generation

In light propagation through optical fiber, the important thing to note is that, if dispersion is anomalous and nonlinearity is present, the frequency chirps created by dispersion and SPM are opposite in nature. Hence, it is possible for the two chirps to cancel, giving a chirp-free pulse, when GVD and SPM exactly balance each other, i.e. $L_D = L_{\mathrm{NL}}$. In this case, neither the pulse nor its spectrum broadens. This undistorted pulse is called a soliton [7, 21]. Let us consider the propagation equation with Kerr-like nonlinearity and second-order dispersion; equation (4.40) reduces to [7, 21]

$$\frac{\partial U}{\partial z} + \frac{i\beta_2}{2}\frac{\partial^2 U}{\partial T^2} = i\gamma(|U|^2 U). \tag{4.57}$$

The fundamental soliton solution to this equation can be directly obtained by assuming a shape-preserving solution of the form $U(z, T) = V(T)\exp[i\phi(z, T)]$. The result is [21]

$$U(z, T) = \sqrt{(}P_0)\mathrm{sech}\left(\frac{T}{T_0}\right)\exp\left[\frac{i|\beta_2|}{2T_0^2}z\right], \tag{4.58}$$

with $N = 1$, where

$$N^2 = \frac{\gamma P_0 T_0^2}{|\beta_2|}. \tag{4.59}$$

Here, N is the soliton order. In the context of optical fibers, solution (4.58) indicates that if a hyperbolic secant pulse of $N = 1$ in equation (4.59) with pulse width T_0 and peak power P_0 is launched inside an ideal lossless fiber, the pulse propagates without a change in shape for arbitrarily long distances. This is the unique feature of the fundamental soliton which makes it attractive for optical communications systems. The peak power P_0 required to support the fundamental soliton is obtained from equation (4.59) by setting $N = 1$ and is given by

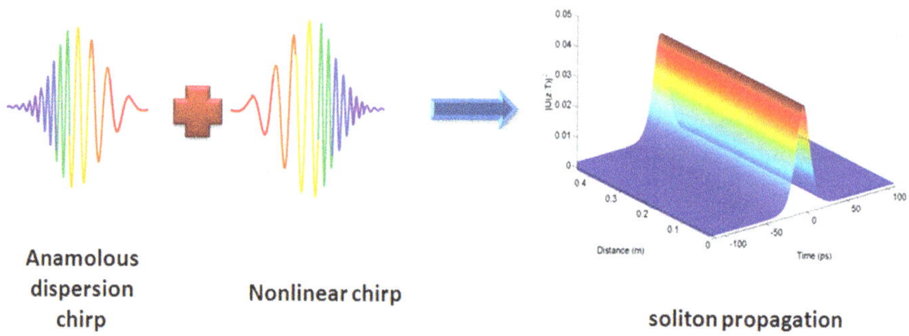

Figure 4.4. Formation of a soliton by balancing the chirps induced by SPM and anomalous GVD.

$$P_0 = \frac{|\beta_2|}{\gamma T_0^2}. \tag{4.60}$$

From the above relation, the power required to support a fundamental soliton for the parameters $\beta_2 = -0.01$ ps^2 m^{-1}, $\gamma = 27.99$ (Wm)$^{-1}$, and $T_0 = 200$ fs is calculated to be 5.36 W. Single-soliton propagation preserves a hyperbolic secant shape for very long distances. Figure 4.4 shows the fundamental soliton evolution obtained by balancing the chirps induced by anomalous GVD and SPM. For integer values of N larger than one, higher-order solitons are formed, which does not preserve their shape during propagation. Instead, such waves follow a periodic evolution during propagation, in which their shape recovers at multiples of the soliton period defined by $\pi/2L_D$. This basic phenomenon of solitons plays a major role in achieving broadband spectra and ultrashort pulses in supercontinuum generation (SCG) and pulse compression.

4.4.3 Modulational instability

Extensive research has been carried out in the field of pulse propagation in optical fibers [27]. A continuous wave (CW) with a cubic nonlinearity in an anomalous dispersion regime is known to develop instability in the form of small modulations in amplitude or phase in the presence of noise or any other weak perturbation; this is called modulational instability (MI) [7, 19]. MI is an instability mechanism driven by soliton dynamics, which was first proposed by Hasegawa and Brinkman in 1980 [28]. The MI phenomenon has been discovered in fluids [29], nonlinear optics [30], and plasmas [31]. In terms of applications, MI provides a natural means of generating ultrashort pulses at ultrahigh repetition rates, making it potentially useful for the future development of high-speed optical communications systems. As a result, MI has been extensively used in many theoretical and experimental studies for the realization of laser sources for ultrahigh-bit-rate optical transmissions [7, 19].

MI can be interpreted as a four-wave mixing (FWM) process in which phase matching has done through nonlinear and dispersive effects; this results in the exponential growth of Stokes and anti-Stokes sidebands at the expense of two pump photons [32]. In addition to noise-induced MI, it is also possible to initiate this process by adding a counter signal with a frequency separation lying within the gain window. This way of achieving MI by means of a co-propagating signal was proposed by Hasegawa in 1984 and verified experimentally by Tai *et al* [33]. During the breakup of CW and quasi-CW radiation into a train of picosecond and femtosecond pulses in the fiber, higher-order nonlinear effects (such as self-steepening and SRS) and higher-order dispersion effects (such as third-order dispersion (TOD)) and fourth-order dispersion (FOD) should also be taken into account [7]. The influences of SPM, higher-order nonlinear effects and higher-order dispersion effects on MI in the anomalous dispersion regime have been described in [7] and the following conclusion was drawn: the instability conditions that govern the generation of ultrashort pulses in the anomalous dispersion regime are unaffected by the presence or absence of the TOD coefficient. The effect of SRS on MI is comparatively small for small values of the perturbation frequency, so the GVD and SPM terms dominate, whereas the gain spectrum increases linearly for comparatively large values of the perturbation frequency, with the result that the MI region is widened due to SRS. Moreover, the self-steepening effect reduces the maximum gain and bandwidth.

4.5 Importance of optical solitons

Solitons are localized solitary waves which have very special properties; for example, (a) they propagate at constant speed without changing their shape; (b) they are extremely stable in the presence of perturbations, and in particular to collisions with small-amplitude linear waves; and (c) they are even stable with respect to collisions with other solitons. When two solitons collide, they pass through each other and emerge with their original speed and shape after the interaction. It should be noted that this interaction is not the simple superposition of the two waves. Moreover, after the collision, the trajectories of the two waves are shifted with respect to their trajectories without the collision. In other words, the outcome of the collision of two solitons is only a simple phase shift of each wave.

Apart from the usage of solitons in communications, they find application in the construction of optical switches. The advantage of using solitons is that they do not change their shapes, even on interaction with other pulses. In these switches, the propagation of one optical pulse affects the other, i.e. the 'signal' pulse is affected by the 'control' pulse. Here, a soliton behaves like the control pulse opening a gate for a signal pulse, so as to allow it to pass through. Photonic logic gates operate on this principle; thus, logic gates can also be realized using solitons.

Other very important applications of optical solitons include the generation of ultrashort pulses through SCG and the pulse-compression technique. Ultrashort pulses are needed in many branches of science which use nanosecond and picosecond pulses. Optical solitons help in the generation of ultrashort pulses through the use of the pulse-compression technique in fiber. The roles of solitons in SCG and pulse compression are discussed in detail in the following chapters.

4.6 Why solitons in photonic crystal fiber?

As a result of the spectacular features of optical solitons in the field of fiber-optic communications, soliton-based applications form one of the hottest research fields and deserve the interest of researchers across the globe [34, 35]. Because of their potential applications, soliton-based applications will become a reality in the near future. In particular, the use of solitons in fiber plays a significant role in the field of SCG and pulse compression [36, 37]. In conventional optical fibers, the generation of solitons in a relatively short length of fiber is practically impossible, since it requires hundreds of kilometers of fiber to counteract dispersion with SPM. However, the invention of photonic crystal fiber (PCF) has helped in the generation of solitons in short lengths of fiber, on the order of centimeters [34]. This is purely because of the second-order dispersion (SOD) coefficient in PCF, which is a hundred orders of magnitude greater than that of the conventional telecommunications fiber. Because of the large value of the dispersion coefficient, the interaction length is reduced drastically, thus enabling pulse compression at length scales of centimeters [36]. In addition, with an appropriate PCF structural design, the zero-GVD wavelength can even be shifted into the visible regime. Solitonic behavior has been investigated in different PCFs by changing their designs and dopant materials. For an example, Skryabin predicted the existence of and studied coupled core–surface solitons in hollow-core PCFs [38]. Podlipensky *et al* experimentally studied the formation and propagation of stable bound-soliton pairs in a highly nonlinear PCF [17]. These bound pairs occurred at a particular power as the consequence of high-order soliton fission. In 2004, Ferrando *et al* proposed a novel PCF-based way of generating vortex solitons, in which the presence of defects played a crucial role [15]. They also demonstrated that large-scale PCF's can support stable nonlinear localized solutions of spatial solitons, which are characterized by a discrete symmetry and differ from the so-called discrete spatial solitons [39]. In addition, Salgueiro *et al* studied spatial vector solitons in a PCF made of a material with focusing Kerr nonlinearity [16]. Analysis of the components showed that two-component localized nonlinear waves consist of two mutually trapped components confined by the PCF and the self-induced nonlinear refractive indices which bifurcate from the corresponding scalar solitons. Pure rotational stimulated Raman scattering spectra containing nine strong spectral components were generated by Benabid *et al* from a ~11 m long hollow PCF filled with hydrogen and pumped with nanosecond pulses that had energies of around 100–300 nJ [40].

Recent measurements of the nonlinear index coefficient n_2 in silica fibers yielded values in the range of 2.2–3.4 $\times$ 10^{-20} m^2 W^{-1}, depending on the core composition of the fiber [7]. One of the best advantages of PCF, apart from being 'endlessly single-mode,' is its enhanced nonlinearity [55]. The enhanced nonlinearity of this fiber can effectively be tailored by suitable PCF structural design, while its strong light-field confinement is due to the high refractive index step between its core and cladding. The high degree of light-field confinement, on the other hand, radically enhances the whole catalog of nonlinear optical processes and allows new nonlinear optical phenomena to be observed [37, 41]. This is made possible by two important characteristics of PCF: a small spot size and extremely low loss. Recently, Travers explained the generation of blue solitary waves through the MI of a continuous infrared wave, which led to red-shifting solitons and blue-shifting dispersive waves [18]. Yiou *et al* showed that high-efficiency stimulated Raman scattering can be obtained using hollow-core PCF whose core was filled with a low-refractive-index nonlinear liquid [42]. PCFs outperform conventional fiber in a variety of applications, including SCG, pulse compression, optical switching, fiber lasers, parametric amplifiers, modulational instability (MI), and others. Despite substantial research into PCF for a large number of applications, the generation of broadband sources and ultrashort pulses using SCG and pulse-compression techniques has a wide range of potential applications in today's extremely demanding technological world.

Achieving efficient soliton propagation in PCF with a single mode, low loss, and a high nonlinear coefficient is a critical challenge in optical soliton transmission lines. In general, there are two proven methods for improving nonlinearity and dispersion in fibers. The first technique is to simulate a PCF with a large air hole size, which results in strong nonlinearity. In 2006, Skryabin *et al* used analytical and numerical methods to demonstrate pulse propagation while accounting for the quantum description of the Raman transition and for optical Kerr nonlinearity and obtained a multi-parameter family of non-topological solitary-wave solutions [38]. Recently, Laegsgaard investigated the formation of solitons by the compression of linearly chirped pulses in hollow-core photonic bandgap fibers (PBGFs) using numerical simulation [43]. They studied the properties of these solitons and reported a transition from stable to unstable regimes of propagation. Gorbach *et al* reported the numerical investigation of several effects accompanying the propagation of femtosecond pulses in air-core PCFs [44]. They found that if the air contribution to the material response dominates over that of silica, then soliton pulses with durations less than or close to 100 fs suffer from strong energy losses into non-solitonic radiation. They also studied soliton behavior in different PCF structures with different doping materials. Very recently, Conti *et al* predicted the existence of a novel class of temporally localized waves propagating inside PCFs with a central core filled by nonlinear liquids that had slow reorientational nonlinearity [45]. However, when a large air hole was used, single-mode propagation could not be maintained. The second approach employs non-silica technology, such as SF6, TF10, Cs_2, nitrobenzene, and others, which helps in attaining nonlinearity for soliton propagation with low-input-energy pulses, and is emerging as one of the most promising possibilities in the development of PCFs.

Similarly, various physical phenomena have been observed in different PCF structures. For instance, Skryabin and his group investigated solitons with variations of both dispersion and nonlinear parameters by employing the exact solution of a nonlinear equation with variable PCF structural designs [39, 46, 47]. In addition, Raman-induced soliton pulse propagation combined with higher-order nonlinear effects has been predicted in PCF [48]. As a result of employing femtosecond solitons in PCF, energy exchange between solitons and polarization instability were demonstrated in [46]. Travers *et al* showed that the trapping of dispersive waves by solitons is significantly enhanced in tapered optical fibers as compared with non-tapered fibers [49]. This opens up the possibility of trapping dispersive waves using solitons in fibers or waveguides made from materials with negligible Raman self-scattering. Fiorentino *et al* experimentally demonstrated the generation of amplitude-squeezed light by means of solitons in a PCF [50]. The results showed that due to the high nonlinearity of the microstructure fiber, squeezing was obtained for smaller peak powers and shorter fiber lengths. Apart from the form of solitons in PCF, the influence of frequency shifts on pulse stability has been analyzed in detail [51]. Likewise, Li *et al* used an MI method to measure the polarization-dependent dispersion of a PCF [52]. They experimentally demonstrated the impact of polarization-dependent dispersion on CW SCG. In addition, Zheltikov demonstrated that MI in PCF supports the generation of pulse sequences with tunable time intervals between pulses, offering attractive solutions for optical telecommunications technologies [53]. For example, such PCF pulse shapers are ideally suited for coherence-controlled nonlinear Raman scattering, including single-beam coherent anti-Stokes Raman scattering microspectroscopy. In 2006, Tonello *et al* experimentally observed the simultaneous generation of polarization and intermodal noise-seeded parametric amplification through MI by injecting quasi-CW intense linearly polarized pump pulses. Furthermore, by shifting the pump wavelength from 532 to 625nm, they observed a shift of polarization sidebands from 3 to 8 THz, whereas the intermodal sidebands shifted from 33 to 63 THz [54]. Numerical simulations of the onset phase of CW SCG from MI by Dudley *et al* showed that the structure of the developing field can be interpreted in terms of the properties of Akhmediev breathers [20].

References

[1] Russell J S 1844 *Report of the 14th meeting of the British Association for the Advancement of Science, York, 1844* (London: John Murray)

[2] Boussinesq J 1871 Théorie de l'intumescence liquide appelée onde solitaire ou de translation se propageant dans un canal rectangulaire *CR Acad. Sci. Paris* **72** 1871

[3] Korteweg D J and De Vries G 1895 XLI. On the change of form of long waves advancing in a rectangular canal, and on a new type of long stationary waves *Lond. Edinb. Dublin Philos. Mag. J. Sci.* **39** 422–43

[4] Zabusky N J and Kruskal M D 1965 Interaction of "solitons" in a collisionless plasma and the recurrence of initial states *Phys. Rev. Lett.* **15** 240

[5] Hasegawa A and Tappert F 1973 Transmission of stationary nonlinear optical pulses in dispersive dielectric fibers *Appl. Phys. Lett.* **23** 142–4

[6] Mollenauer L F, Stolen R H and Gordon J P 1980 Experimental observation of picosecond pulse narrowing and solitons in optical fibers *Phys. Rev. Lett.* **45** 1095

[7] Agrawal G P 2007 *Nonlinear Fiber Optics* 4 (Cambridge, MA: Academic Press)

[8] Kumar R, Malik H K and Kawata S 2011 Soliton reflection in a plasma with trapped electrons: the effect of dust concentration *Physica* D **240** 310–6

[9] Ustinov A B, Demidov V E, Kondrashov A V, Kalinikos B A and Demokritov S O 2011 Observation of the chaotic spin-wave soliton trains in magnetic films *Phys. Rev. Lett.* **106** 017201

[10] Atanasov V and Omar Y 2010 Quanta of local conformational change: conformons in α-helical proteins *New J. Phys.* **12** 055003
[11] Antonelli P and Sparber C 2011 Existence of solitary waves in dipolar quantum gases *Physica* D **240** 426–31
[12] Hori T, Nishizawa N, Goto T and Yoshida M 2004 Experimental and numerical analysis of widely broadened supercontinuum generation in highly nonlinear dispersion-shifted fiber with a femtosecond pulse *J. Opt. Soc. Am.* B **21** 1969–80
[13] Warn-Varnas A, Chin-Bing S A, King D B, Hawkins J and Lamb K 2009 Effects on acoustics caused by ocean solitons, part A: Oceanography *Nonlinear Anal. Theory Methods Appl.* **71** e1807–17
[14] Oscoz A, Mediavilla E, Serra-Ricart M, Diaz-Alonso J and Rubiera-Garcia D 2008 Soliton solutions in relativistic field theories and gravitation *Eur. Astron. Soc. Publ. Ser.* **30** 193–6
[15] Ferrando A, Zacarés M, de Córdoba P F, Binosi D and Monsoriu J A 2004 Vortex solitons in photonic crystal fibers *Opt. Express* **12** 817–22
[16] Salgueiro J R, Kivshar Y S, Pelinovsky D E, Simón V and Michinel H 2005 Spatial vector solitons in nonlinear photonic crystal fibers *Stud. Appl. Math.* **115** 157–71
[17] Podlipensky A, Szarniak P, Joly N Y, Poulton C G and Russell P St J 2007 Bound soliton pairs in photonic crystal fiber *Opt. Express* **15** 1653–62
[18] Travers J C 2009 Blue solitary waves from infrared continuous wave pumping of optical fibers *Opt. Express* **17** 1502–7
[19] Agrawal G 2001 *Applications of Nonlinear Fiber Optics* (Amsterdam: Elsevier)
[20] Dudley J M, Genty G, Dias F, Kibler B and Akhmediev N 2009 Modulation instability, akhmediev breathers and continuous wave supercontinuum generation *Opt. Express* **17** 21497–508
[21] Pal B P 2010 Guided Wave Optical Components and Devices: Basics *Technology, and Applications* (New York: Academic)
[22] Taha T R and Ablowitz M J 1988 Analytical and numerical aspects of certain nonlinear evolution equations. IV. Numerical, modified Korteweg-de Vries equation *J. Comput. Phys.* **77** 540–8
[23] Ablowitz M J and Ladik J F 1976 A nonlinear difference scheme and inverse scattering *Stud. Appl. Math.* **55** 213–29
[24] Ablowitz M J and Ladik J F 1977 On the solution of a class of nonlinear partial difference equations *Stud. Appl. Math.* **57** 1–12
[25] John Senior M 1992 *Optical Fiber Communications Principles and Practice* (Englewood Cliffs, NJ: Prentice-Hall)
[26] Keiser G 2003 *Optical Communications Essentials* (New York: McGraw-Hill)
[27] Dudley J M and Taylor J R 2010 *Supercontinuum Generation in Optical Fibers* (Cambridge: Cambridge University Press)
[28] Hasegawa A and Brinkman W 1980 Tunable coherent IR and FIR sources utilizing modulational instability *IEEE J. Quantum Electron.* **16** 694–7
[29] Nakazawa M, Suzuki K, Kubota H and Haus H A 1989 High-order solitons and the modulational instability *Phys. Rev.* A **39** 5768
[30] Ostrovskii L A 1967 Propagation of wave packets and space-time self-focusing in a nonlinear medium *Sov. Phys. JETP* **24** 797–800 http://www.jetp.ras.ru/cgi-bin/e/index/e/24/4/p797?a=list
[31] Karpman V I 1967 Self-modulation of nonlinear plane waves in dispersive media *Sov. J. Exp. Theor. Phys. Lett.* **6** 277 http://jetpletters.ru/ps/1672/article_25503.shtml

[32] Kalithasan B, Porsezian K, Dinda P T and Malomed B A 2009 Modulational instability and generation of self-induced transparency solitons in resonant optical fibers *J. Opt. A: Pure Appl. Opt.* **11** 045205
[33] Tai K, Hasegawa A and Tomita A 1986 Observation of modulational instability in optical fibers *Phys. Rev. Lett.* **56** 135
[34] Raja R V J, Porsezian K, Varshney S K and Sivabalan S 2010 Modeling photonic crystal fiber for efficient soliton pulse propagation at 850 nm *Opt. Commun.* **283** 5000–6
[35] Duling I N (ed) 1995 *Compact Sources of Ultrashort Pulses* (Cambridge: Cambridge University Press)
[36] Raja R V J, Senthilnathan K, Porsezian K and Nakkeeran K 2010 Efficient pulse compression using tapered photonic crystal fiber at 850 nm *IEEE J. Quantum Electron.* **46** 1795–803
[37] Raja R V J, Husakou A, Hermann J and Porsezian K 2010 Supercontinuum generation in liquid-filled photonic crystal fiber with slow nonlinear response *J. Opt. Soc. Am.* B **27** 1763–8
[38] Skryabin D V 2004 Coupled core-surface solitons in photonic crystal fibers *Opt. Express* **12** 4841–6
[39] Ferrando A, Zacarés M, de Córdoba P F, Binosi D and Monsoriu J A 2003 Spatial soliton formation in photonic crystal fibers *Opt. Express* **11** 452–9
[40] Benabid F, Antonopoulos G, Knight J C and Russell P St J 2005 Stokes amplification regimes in quasi-cw pumped hydrogen-filled hollow-core photonic crystal fiber *Phys. Rev. Lett.* **95** 213903
[41] Raja R V J, Porsezian K and Nithyanandan K 2010 Modulational-instability-induced supercontinuum generation with saturable nonlinear response *Phys. Rev.* A **82** 013825
[42] Yiou S *et al* 2005 Stimulated Raman scattering in an ethanol core microstructured optical fiber *Opt. Express* **13** 4786–91
[43] Lægsgaard J 2009 Soliton formation in hollow-core photonic bandgap fibers *Appl. Phys.* B **95** 293–300
[44] Gorbach A V and Skryabin D V 2008 Soliton self-frequency shift, non-solitonic radiation and self-induced transparency in air-core fibers *Opt. Express* **16** 4858–65
[45] Conti C, Schmidt M A, Russell P St J and Biancalana F 2010 Highly noninstantaneous solitons in liquid-core photonic crystal fibers *Phys. Rev. Lett.* **105** 263902
[46] Luan F, Yulin A V, Knight J C and Skryabin D V 2006 Polarization instability of solitons in photonic crystal fibers *Opt. Express* **14** 6550–6
[47] Skryabin D V, Luan F, Knight J C and Russell P St J 2003 Soliton self-frequency shift cancellation in photonic crystal fibers *Science* **301** 1705–8
[48] Efimov A and Taylor A J 2005 Spectral-temporal dynamics of ultrashort Raman solitons and their role in third-harmonic generation in photonic crystal fibers *Appl. Phys.* B **80** 721–5
[49] Travers J C and Taylor J R 2009 Soliton trapping of dispersive waves in tapered optical fibers *Opt. Lett.* **34** 115–7
[50] Fiorentino M, Sharping J E, Kumar P, Porzio A and Windeler R S 2002 Soliton squeezing in microstructure fiber *Opt. Lett.* **27** 649–51
[51] Hu M-L, Li Y-F, Chai L, Xing Q, Doronina L V, Ivanov A A, Wang C-Y and Zheltikov A M 2008 Two-dimensional coherent superposition of blue-shifted signals from an array of highly nonlinear waveguiding wires in a photonic-crystal fiber *Opt. Express* **16** 11176–81
[52] Li X, Wei Z, Yi-Dong H and Jiang-De P 2008 Polarization dependent dispersion and its impact on optical parametric process in high nonlinear microstructure fibre *Chin. Phys.* B **17** 995

[53] Zheltikov A M 2008 Pulse shaping by modulation instability in a photonic-crystal fiber for coherence control and single-beam coherent anti-stokes Raman-scattering microspectroscopy *Laser Phys.* **18** 1465–78

[54] Tonello A, Pitois S, Wabnitz S, Millot G, Martynkien T, Urbanczyk W, Wojcik J, Locatelli A, Conforti M and De Angelis C 2006 Frequency tunable polarization and intermodal modulation instability in high birefringence holey fiber *Opt. Express* **14** 397–404

[55] Zhang X, He M, Chang M, Chen H, Chen N, Qi N, Yuan M and Qin X 2018 Dual-cladding high-birefringence and high-nonlinearity photonic crystal fiber with As_2S_3 core *Opt. Commun.* **410** 396–402

IOP Publishing

Fiber Optic Pulse Compression
Numerical techniques and applications with MATLAB®
R Vasantha Jayakantha Raja and A Esther Lidiya

Chapter 5

Conventional compression schemes

This chapter explores existing conventional soliton compression techniques, such as higher-order soliton compression and adiabatic pulse compression, used with PCF to generate ultrashort pulses. It also provides a quality analysis of the compressed pulses.

5.1 Mechanism of pulse compression

Numerous methods for dynamic pulse compression have been proposed that simultaneously broaden the spectrum of the pulse and provide a stable spectral phase from shot to shot. Among these methods, two techniques are widely used to achieve optical pulse compression, namely higher-order soliton compression and adiabatic pulse compression [1, 2].

5.2 Soliton compression

A well-known approach for compressing ultrashort pulses is based on the dynamics of higher-order soliton propagation in optical fiber [3–5]. When a pulse with a sufficiently high peak power propagates in a medium with anomalous group velocity dispersion (GVD), dispersive effects can balance the effects of self-phase modulation (SPM), resulting in the formation of a fundamental soliton. Higher-order solitons are produced at sufficiently high input powers. During their transmission, these higher-order solitons undergo oscillatory variation in the pulse's temporal structure and spectrum [6, 7]. The pulse is significantly compressed at the start of each period, depending on the soliton sequence. With an appropriate choice of fiber length, the output pulse width can be significantly shorter than the input pulse width.

5.2.1 Second-order soliton compression

The letter N is used to represent the soliton order; for example, the value $N = 2$ denotes a second-order soliton. As stated in the preceding section, the soliton order N is proportional to the power of the launched optical pulse. When $N = 2$, the optical pulse begins to evolve in a periodic way, which is not the case with a fundamental soliton. As previously mentioned, the influence of SPM produces positive chirp, whereas the effect of β_2 (anomalous dispersion) produces negative chirp. At higher intensities, the proportion of positive chirp caused by SPM is more than the amount of negative chirp produced by β_2, and therefore the positive chirp cannot be entirely canceled out. As a result, the pulse acquires a positive chirp. For a positively chirped pulse, the impact of GVD ($D > 0$) is to compress the pulse by driving the energy associated with its spectral components toward the center. When they are moved past the centre, the pulse expands again. Figures 5.1(a) and (b) show the periodic evolution of an optical pulse in the temporal and spectral domains.

During the evolution of the pulse in the time domain, one can clearly see the contraction and recovery in the pulse shape. The distance over which the pulse regains its original pulse shape is called the soliton period and denoted by z_0. The analytical solution for z_0 is [8]:

$$z_0 = \frac{\pi}{2} L_D = \frac{\pi}{2} \frac{T_0^2}{\beta_2} \approx \frac{T_{\text{FWHM}}^2}{2|\beta_2|}. \tag{5.1}$$

For $N > 1$, the amount of positive chirp created by SPM is too big to be canceled out by the amount of negative chirp generated by GVD. As a result, it accumulates and grows until it reaches the distance $z_0/2$, as shown in figure 5.1(a), where the shifted energy due to positive dispersion coincides at the pulse center. When the shifted energy passes through the pulse center, the chirp's sign changes from positive to negative. For a negatively chirped pulse, GVD (D>0) has the effect of broadening the pulse. The pulse begins to re-broaden at a distance of $z_0/2$. Because the positive chirp produced by SPM swiftly cancels the negative chirp of the pulse, the quantity of negative chirp is reduced and the spread of the pulse is slowed. At a distance of z_0, where the pulse stabilizes at its original width, the chirp becomes zero. The procedure is then repeated. The pulse spectrum broadens in the spectral domain as the pulse compresses. As a result, the compressed pulse may be obtained near the end of a fiber that has a length of $z_0/2$.

5.2.2 Third-order soliton compression

As the soliton order grows, the interaction between SPM and GVD becomes more complex. Figures 5.2(a), (b), and (c) depict the periodic evolution of an $N = 3$ soliton over one soliton period in the temporal and spectral domains. As the soliton propagates along the fiber, the evolution of the soliton is symmetric around $z_0/2$ and repeats at a distance of every z_0. The pulse contracts with increasing peak power during the early stages of evolution. It then divided into two pulses near $z_0/2$. After $z = z_0/2$, the process reverses. It combines once again and returns to its original form. Pulse compression is a dominant feature in the early stages of evolution (up to $z = 11.66$ m). Since SPM always produces a positive chirp during propagation through the fiber, in the anomalous GVD zone, the pulse compresses like a positively pre-chirped pulse. Because the positive chirp is linear towards the center of the pulse, only this portion of the pulse is compressed. The rest of the pulse has

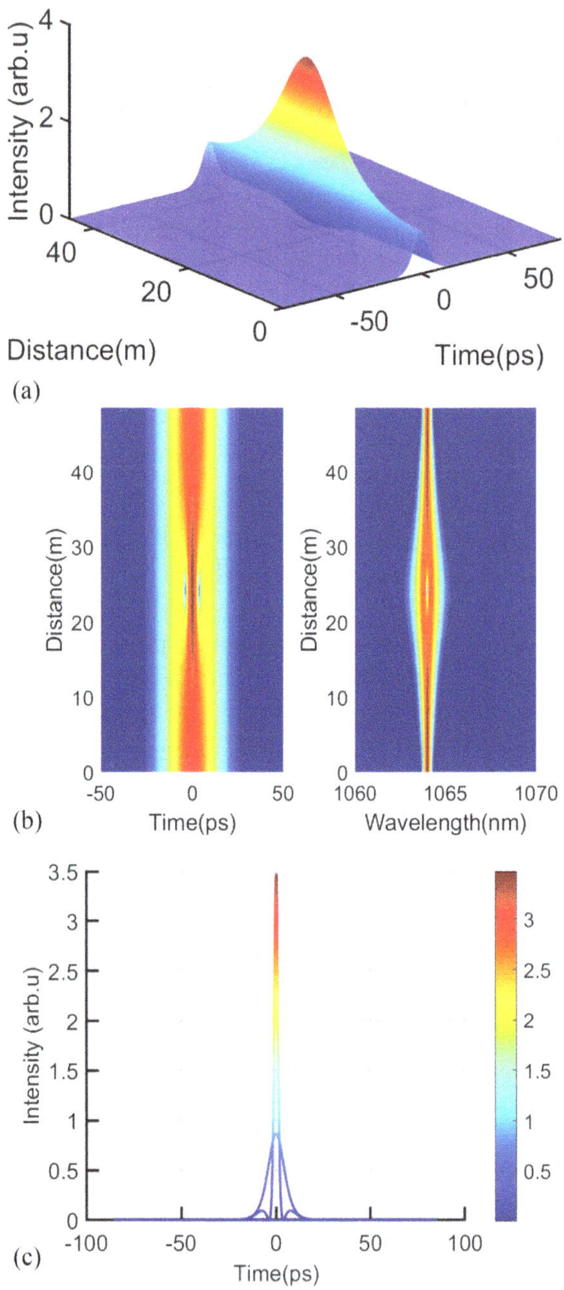

Figure 5.1. Compression of a soliton pulse while injecting the power corresponding to a second-order soliton: (a) temporal evolution, (b) two-dimensional representation of a second-order soliton in the temporal and spectral domains, (c) final compressed pulse profile at 24.2 m.

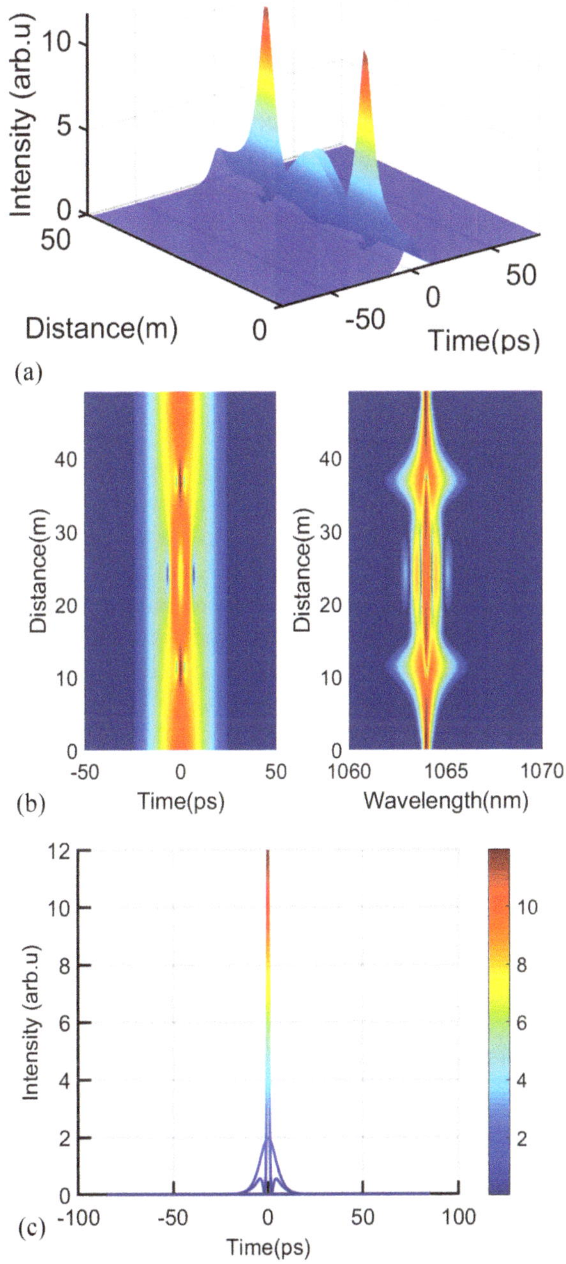

Figure 5.2. Compression of a soliton pulse while injecting the power corresponding to a third-order soliton: (a) evolution in the time domain, (b) two-dimensional representation of soliton compression in the temporal and spectral domains, (c) compression of a third-order soliton at 11.66 m.

negative chirp; hence, it disperses and creates side lobes resembling a pulse shape at $z = 50\,\text{m}$. The pulse spectrum of the first stage clearly exhibits the characteristic SPM-induced spectral broadening at $z = 18\,\text{m}$. The pulse then begins to separate at $z = 25\,\text{m}$. As a result, the compressed pulse is detected near a length of 11.66 m in the first pulse spectrum.

Since the pulse wings are not as completely compressed as the central component, the compressed output pulse, which is just a small percentage of the total energy, rests on a broad pedestal [9]. The disadvantage of this approach is that the compressed pulse quality is relatively poor because of the accumulation of a considerable percentage of the pulse energy in the wide pedestal rather than the compressed spike. For a pulse-compression factor of 60, the pedestal component contains up to 80% of the pulse's energy [10]. Because the pedestals produce intersymbol interference, these compressed pulses cannot be employed as long-distance information carriers.

5.3 Quality analysis

The effectiveness of the modeled pulse-compression system can be quantified by few analytical parameters, namely, the compression factor (F_C), the pedestal energy (PE), and the quality factor (Q_F). The major factors that determine the efficiency of the designed compressor are the compression factor and the pedestal energy [11]. In order to obtain better pulse shape quality, the optimal optical fiber for pulse compression should simultaneously produce a greater compression factor and a reduced pedestal energy in propagating pulses.

5.3.1 Compression factor

In general, the compression factor (F_C) is defined as the ratio of the full width at half maximum (FWHM) of the input pulse to that of the output compressed pulse's FWHM, which is given by

$$F_c = \frac{T_{\text{FWHM, in}}}{T_{\text{FWHM, out}}} \tag{5.2}$$

where $T_{\text{FWHM, in}}$ and $T_{\text{FWHM, out}}$ are the FWHM of the input pump pulse and the output compressed pulse, respectively.

5.3.2 Pedestal energy

The pulse develops side lobes around the central compressed pulse during propagation along the fiber length. The energy wasted in the wings of the compressed pulse is defined as pedestal energy. Mathematically, the pedestal energy is given by

$$E_{\text{pedestal}} = \frac{|E_{\text{out}} - E_{\text{sech}}|}{E_{\text{out}}} \times 100\% \tag{5.3}$$

E_{pedestal} is an important parameter that provides information about the quality of the pulse and quantifies the percentage of the total input energy that is contained in the pedestal of the compressed pulse. In equation (5.3), the energy of the hyperbolic secant pulse is given by

$$E_{\text{sech}} = 2P_{\text{peak}}\frac{T_{\text{FWHM}}}{2\log(1+\sqrt{2})}. \tag{5.4}$$

A pedestal energy of more than 40% means that the quality of the pulse shape is seriously compromised. In general, a large pedestal energy signifies that the quality of the compressed pulse is poor. If there is no pedestal in the output pulse, the quality factor Q_c is unity. Thus, the preferred choice for pulse compression is fiber that yields the greatest pulse shape quality along with the highest compression factor.

5.3.3 Quality factor

In addition, the quality of the compressed pulse is analyzed using the quality factor (Q_c), which is defined as the fraction of energy contained in the output compressed pulse with respect to that of the input pulse, that is

$$Q_c = \frac{E_{\text{out}}}{E_{\text{in}}} \tag{5.5}$$

where $E_{\text{in}}$and E_{out} are the energies of the seed pulse and the compressed pulse at the output, respectively. The quality factor determines the energy sustained by the proposed pulse-compression system. For an effective compression system, the quality factor should be unity.

5.4 Adiabatic compression

Adiabatic soliton compression is a technique for the temporal compression of ultrashort pulses in a fiber. To realize high-quality pulse compression, the pedestal must be minimized, because interaction between the pedestal and the compressed spike occurs upon further propagation, leading to undesirable periodic pulse reshaping [12, 13]. For this reason, the compression of fundamental solitons with no pedestal component has been widely investigated. Chernikov *et al* [14] and Blow *et al* [15] have shown that such soliton compression can be achieved using the adiabatic amplification of solitons in fibers. To investigate pulse compression achieved through the adiabatic process, the modified nonlinear Schrödinger equation (MNLSE) is numerically solved using the SSFM with the initial envelope of a sech-shaped pulse at $z = 0$. Nonlinear pulse propagation in the tapered PCF is described by the following MNLSE, which includes varying dispersion and the nonlinear coefficient [2]:

$$\frac{\partial U}{\partial z} - \sum_{n=2}^{3}\beta_n(z)\frac{i^{n+1}}{n!}\frac{\partial^n U}{\partial t^n} = i\gamma(z)\left[1+\frac{i}{\omega_0}\frac{\partial}{\partial t}\right]\left[U(z,t)\int_0^{\infty}R(t')|U(z,t-t')|^2dt'\right] \tag{5.6}$$

In solving equation (5.6), in order to observe the conservation of energy during pulse propagation, the pulse parameters and fiber parameters have to satisfy the following equation:

$$T_0 = \frac{2N^2|\beta_2|}{E_s\gamma}, \tag{5.7}$$

where E_s is the energy of the soliton. Thus, the pulse duration can be reduced by decreasing the dispersion while keeping the pulse energy constant. This technique is known as the adiabatic compression technique. The principle is as follows: for a fundamental soliton pulse in a fiber, the product of the pulse energy and the pulse duration is proportional to the ratio of the dispersion (β_2) and nonlinearity (γ) parameters of the fiber. Hence, the use of fibers with variable dispersion is viewed as an effective method for controlling soliton propagation in the generation of low-pedestal ultrashort pulses. Thus, very high-quality pulse compression is possible through adiabatic pulse compression, and the input power requirements are significantly lower than for soliton-effect compression. In particular, dispersion-decreasing fibers (DDFs) have been recognized to be very useful for high-quality, stable, polarization-insensitive, adiabatic soliton pulse compression [16–18]. In order to obtain significant pulse compression through a dispersion-decreasing fiber, the following conditions need to be satisfied:

- The initial pulses must fulfill the soliton condition at the input fiber end.
- The fiber dispersion must be varied slowly enough to allow adiabatic adaptation of the pulse to the fiber parameters. Otherwise, the pulse shape can become distorted.

Very short pulses have been already achieved through experimental and numerical simulation using adiabatic pulse compression in PCF at different wavelengths [19, 20]. Recently, Travers *et al* investigated adiabatic pulse compression at 1.06 μm using dispersion and variable-nonlinearity PCF [21]. They experimentally proved that one can achieve a high compression factor using dispersion-decreasing and nonlinearity-increasing PCF for adiabatic pulse compression. Hence, in order to explain the adiabatic compression process, a numerical model of a tapered PCF that satisfies the adiabatic condition is depicted here. Recent articles by Dudley *et al* and Richardson *et al* [19, 22] pointed out that one has freedom to choose any path to fabricate a tapered PCF to take advantage of dispersion-decreasing and nonlinearity-increasing profiles. However, we have considered one of the best optimized paths so as to construct a compact compressor using a short length of PCF; this can be realized by choosing low dispersion lengths.

To study adiabatic pulse compression at 1064 nm, we designed a PCF with the maximum possible dispersion to obtain a low dispersion length $L_D(=T_0^2/|\beta_2|)$ within the single-mode regime, where T_0 is the pulse width. The tapered PCF with dispersion-decreasing and nonlinearity-increasing parameters at 1064 nm

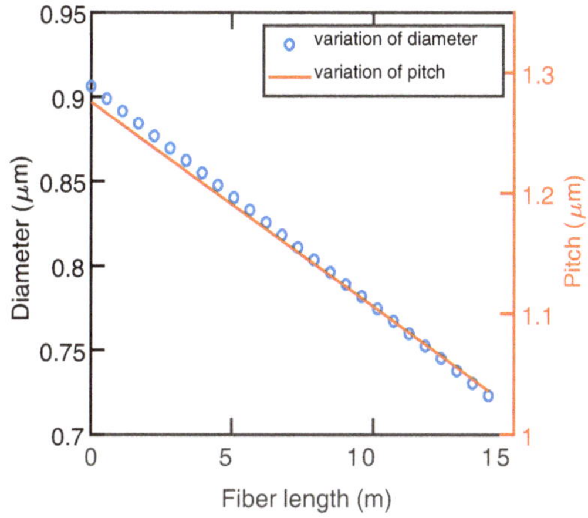

Figure 5.3. The variation in the PCF parameters d and Λ versus propagation distance. The physical parameter profiles for the tapered PCF vary as follows: Λ ranges from 1.27 μm to 1.03 μm and d ranges from 0.91 to 0.73.

was numerically modeled with the help of the finite-element method (FEM). The variation in the fiber parameters such as pitch, air-hole diameter, β_2, β_3, dispersion, and nonlinearity is pictorially represented in figure 5.3. Figure 5.3 illustrates the variation of d and Λ with respect to distance. Both physical parameters d and Λ vary linearly with distance in the tapered PCF. Figure 5.4(a) depicts the variations in β_2 and β_3. In figure 5.4(b), one can observe the decreasing dispersion profile and increasing nonlinearity profile of the modeled tapered PCF at 1064 nm. The maximum GVD value calculated by FEM [23, 24] within the single-mode regime at 1064 nm is 157 ps/nm/km for the PCF parameters d = 0.91 μm and Λ = 1.27 μm. The parameter d varies from 0.91 μm to 0.73 μm and Λ varies from 1.27 μm to 1.03 μm.

The PCF has a length L = 14.09 m, which is four times the dispersion length ($L = 4L_D$), where L_D = 3.52 m for a given pulse width of 1 ps. To investigate the adiabatic compression process in the modeled tapered PCF, equation (5.6) has to be numerically solved using the SSFM with an initial soliton envelope that has a peak power of P_0 = 9.56 W at z = 0 given by $U(0, T) = \sqrt{P_0}$ sech (T). A numerical simulation is carried out for the input pulse at a wavelength of λ = 1064 nm and an FWHM pulse width of 1 ps. Figure 5.5(a) portrays the evolution of the fundamental soliton resulting from the adiabatic compression process. Figure 5.5(b) represents the final compressed pulse profile generated by the adiabatic compression scheme, which graphically illustrates the enhancement in the compression factor of the propagating pulse. Figure 5.6 shows the corresponding spectrogram profile of the input pulse and the adiabatically compressed final output pulse, from which one can simultaneously measure both the pulse width and bandwidth for the PCF length of $4L_D$. The temporal and spectral evolutions of the fundamental soliton along the

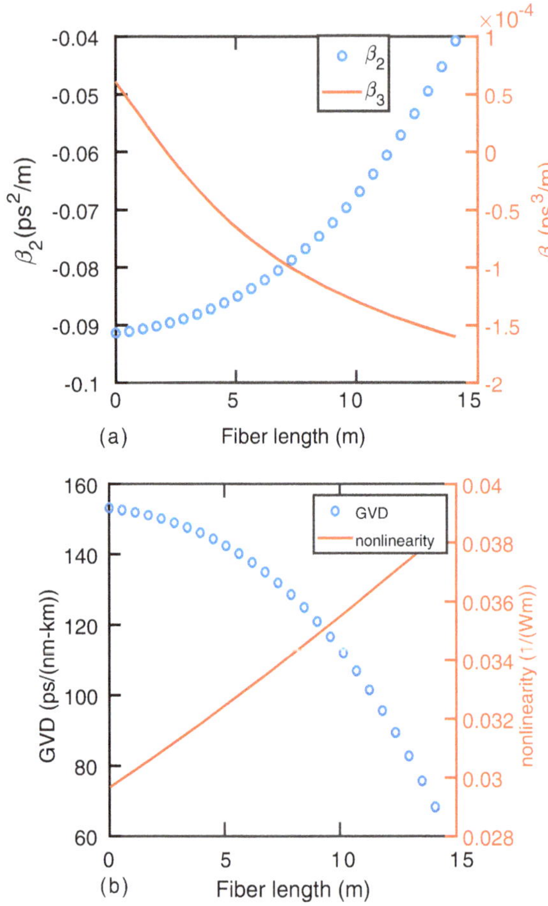

Figure 5.4. (a) Calculated β_2 and β_3 profiles as a function of distance, (b) calculated GVD and nonlinearity as a function of distance.

adiabatically dispersion-decreasing PCF are shown in figures 5.7(a) and (b), which clearly depict the monotonous decrement in the temporal duration of the pulse and the monotonous increment in the spectral width of the pulse. The variation in the peak power and the T_{FWHM} of the pulse along the fiber is represented in figure 5.8. The quality of the compressed pulse can be understood in two ways: by calculating (i) the compression factor and (ii) the pedestal energy. The variations in the compression factor and the pedestal energy along the fiber length are depicted in figure 5.9(a). In this process, the calculated compression factor is 2.17. Although the pulse can be compressed through adiabatic soliton pulse compression, the compression factor is very low at 1064 nm. One can effectively visualize the impact of pedestal on the compressed pulse by plotting the intensity of the pulse on a logarithmic scale. Figure 5.9(b) shows the intensity of the compressed pulse on a logarithmic scale, which allows one to observe the broad pedestal at the bottom of

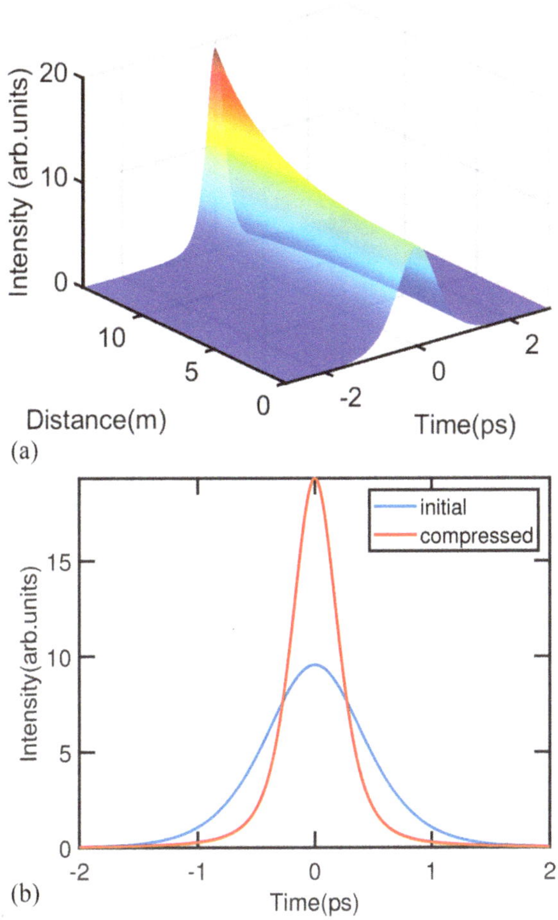

Figure 5.5. (a) Evolution of a fundamental soliton along the fiber, (b) final compressed pulse produced by the adiabatic scheme at 1064 nm for a length of $4L_D$.

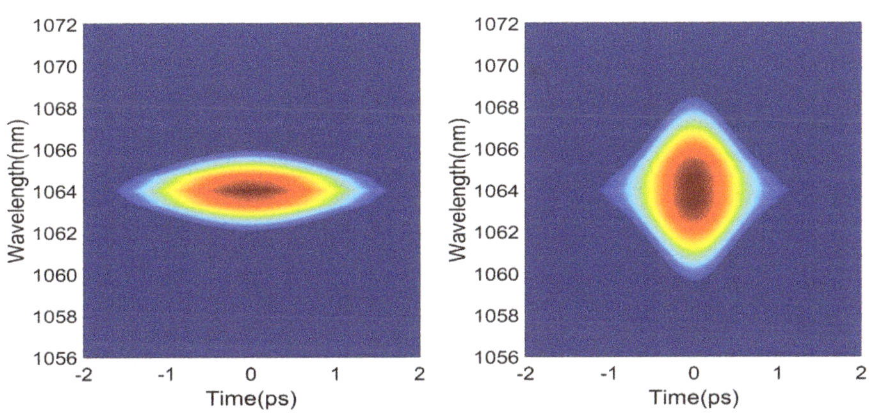

Figure 5.6. Comparison between the spectrogram profile of an initial pulse that has a pulse width of 1 ps FWHM and that of a compressed pulse that has a pulse width of 0.46 ps FWHM.

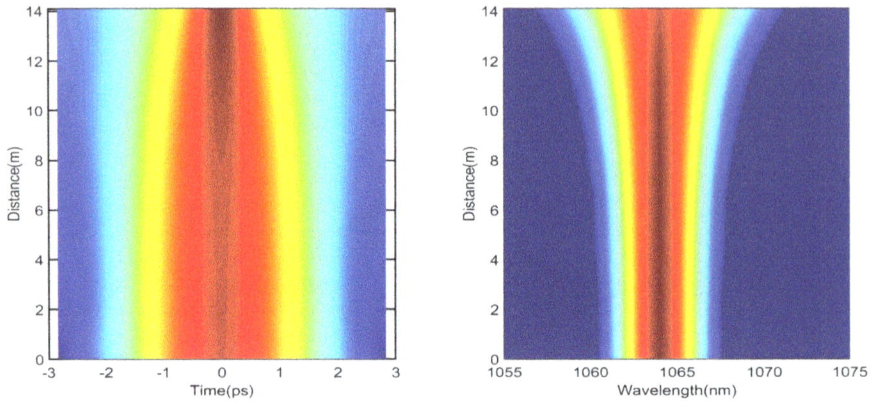

Figure 5.7. Variations in the temporal and spectral profiles of the pulse.

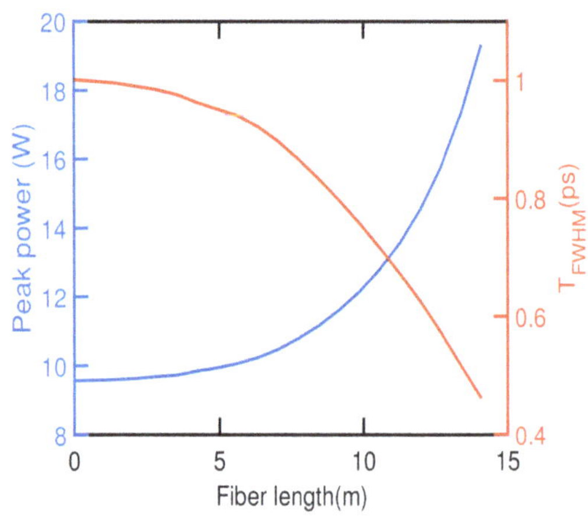

Figure 5.8. Changes in peak power and pulse width along the fiber length.

the pulse; from this, we conclude that the resulting compressed pulse suffers severely from the the presence of a pedestal when this adiabatic compression scheme is used. Pedestal formation is undesirable in all applications, but particularly in optical fiber communications.

5.5 Pulse-parameter equation

In order to analyze the dynamics of a pulse passing through fiber, several analytical and numerical techniques such as the Lagrangian variational method [8], the Hamiltonian method [25], the projection operator method (POM) [26–28], the non-Lagrangian collective variable (CV) approach [29], the CV technique [30], and the moment method [8] have been used. Among these techniques, the

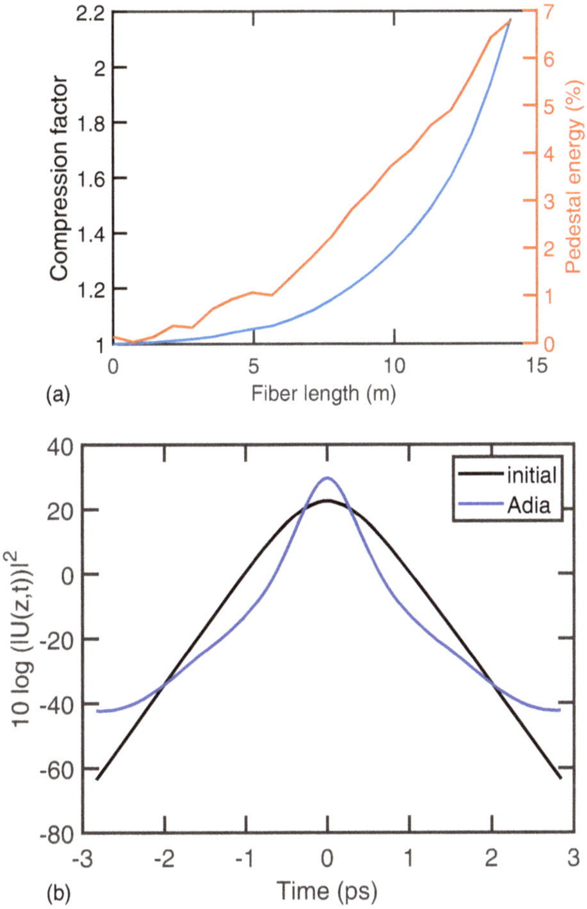

Figure 5.9. (a) Changes in the compression factor and pedestal energy of adiabatic soliton pulse compression for PCF at 1064 nm over four soliton periods, (b) intensity of an adiabatically compressed pulse on a logarithmic scale.

generalized POM is considered to be the most versatile, as it does not require the complex derivation procedure the Lagrangian. Nakkeeran and Wai presented the POM for complex nonlinear partial differential equations as a way of obtaining the ordinary differential equation, which could be obtained using either the Lagrangian variational approach or the bare approximation of the CV theory [27, 28]. Since then, numerous researchers have been inspired to investigate pulse propagation with various linear and nonlinear physical coefficients in single-mode and birefringent fiber using the Gaussian/hyperbolic ansatz [27, 30]. To analyze the evolution of the parameters, one can utilize the generalized POM which can derive the pulse parameter equations from the bare approximation of CV theory [28].

5.6 Projection operator method

The POM basically assumes a predetermined profile for the pulse (for example, Gaussian or hyperbolic secant):

$$\frac{\partial U}{\partial z} + \frac{\alpha}{2}U + \sum_{n=2}^{4}\beta_n(z)\frac{i^{n-1}}{n!}\frac{\partial^n U}{\partial T^n} = i\gamma(z)\left(|U|^2 U + \frac{i}{\omega_0}\frac{\partial(|U|^2 U)}{\partial T} - T_R U\frac{\partial |U|^2}{\partial T}\right). \quad (5.8)$$

Using equation (5.8), one can then derive the dynamical equations (ordinary differential equations) of the pulse parameters (for example, amplitude, width, velocity, and phase). In other words, we can view this as a reduction of the partial differential equation to a set of ordinary differential equations with an appropriate assumption regarding the pulse shape and the necessary pulse parameters. In order to derive the dynamical equation, the POM considers an ansatz $f(x_1, \ldots, x_N, T)$ as a function of the pulse parameters $f(x_1, \ldots, x_N, T)$, where the pulse parameters (also called CVs) are dependent only on z. For complex equations such as the NLSE, a generalized projection operator $P_k = \exp(i\theta)f^*_{xk}$ is introduced, where θ is an arbitrary phase constant. To obtain the CV equation of motion, equation (5.8) is projected in the direction of P_k. By substituting the ansatz function f for U in equation (5.8), multiplying the resulting equation by P_k and integrating with respect to t, one can obtain

$$\begin{aligned}
&\int_{-\infty}^{\infty} \Re[f_z f^*_{x_k}\exp(i\theta)]\, dT - \frac{\alpha}{2}\int_{-\infty}^{\infty} \Re[f f^*_{x_k}\exp(i\theta)]\, dT \\
&- \frac{\beta_2(z)}{2}\int_{-\infty}^{\infty} \Im[f_{tt} f^*_{x_k}\exp(i\theta)]dT - \frac{\beta_3(z)}{6}\int_{-\infty}^{\infty} \Re[f_{ttt} f^*_{x_k}\exp(i\theta)]\, dT \\
&+ \gamma\int_{-\infty}^{\infty} |f|^2\Im[f f^*_{x_k}\exp(i\theta)]\, dT + \frac{\beta_4(z)}{24}\int_{-\infty}^{\infty} \Im[f_{tttt} f^*_{x_k}\exp(i\theta)]\, dT \\
&+ \frac{\gamma}{\omega_0}\int_{-\infty}^{\infty} \Re[(|f|^2 f)_t f^*_{x_k}\exp(i\theta)]dT - \gamma T_R\int_{-\infty}^{\infty} \Im[f(|f|^2)_t f^*_{x_k}\exp(i\theta)]\, dT = 0.
\end{aligned} \quad (5.9)$$

if we substitute $\theta = \pi/2$ into equation (5.9), the equation reduces to

$$\begin{aligned}
&\int_{-\infty}^{\infty} \Im[f_z f^*_{x_k}]dT - \frac{\alpha}{2}\int_{-\infty}^{\infty} \Im[f f^*_{x_k}]dT - \frac{\beta_2(z)}{2}\int_{-\infty}^{\infty} \Re[f_{tt} f^*_{x_k}]\, dT \\
&- \frac{\beta_3(z)}{6}\int_{-\infty}^{\infty} \Im[f_{ttt} f^*_{x_k}]dT + \gamma\int_{-\infty}^{\infty} |f|^2\Re[f f^*_{x_k}]dT + \frac{\beta_4(z)}{24}\int_{-\infty}^{\infty} \Re[f_{tttt} f^*_{x_k}]dT \\
&+ \frac{\gamma}{\omega_0}\int_{-\infty}^{\infty} \Im[(|f|^2 f)_t f^*_{x_k}]dT - \gamma T_R\int_{-\infty}^{\infty} \Re[f(|f|^2)_t f^*_{x_k}]dT = 0.
\end{aligned} \quad (5.10)$$

If we substitute $\theta = 0$ into equation (5.9), the equation reduces to

$$\begin{aligned}
&\int_{-\infty}^{\infty} \Re[f_z f^*_{x_k}]dT - \frac{\alpha}{2}\int_{-\infty}^{\infty} \Re[f f^*_{x_k}]dT - \frac{\beta_2(z)}{2}\int_{-\infty}^{\infty} \Im[f_{tt} f^*_{x_k}]dT \\
&- \frac{\beta_3(z)}{6}\int_{-\infty}^{\infty} \Re[f_{ttt} f^*_{x_k}]dT + \gamma\int_{-\infty}^{\infty} |f|^2\Im[f f^*_{x_k}]dT + \frac{\beta_4(z)}{24}\int_{-\infty}^{\infty} \Im[f_{tttt} f^*_{x_k}]dT \\
&+ \frac{\gamma}{\omega_0}\int_{-\infty}^{\infty} \Re[(|f|^2 f)_t f^*_{x_k}]dT - \gamma T_R\int_{-\infty}^{\infty} \Im[f(|f|^2)_t f^*_{x_k}]dT = 0.
\end{aligned} \quad (5.11)$$

It should be noted that two sets of dynamical equations can be obtained by choosing different projection operator schemes, namely θ equal to zero or θ equal to $\pi/2$. Among these two sets of equations, θ equal to zero is the commonly used case; hence, it is considered here to investigate the impact of fourth-order dispersion (FOD). Here, the hyperbolic secant ansatz has been used to derive the pulse parameter evolution equations:

$$f = x_1 \operatorname{sech}\left(\frac{T - x_2}{x_3}\right) \exp\left(\frac{i x_4 (T - x_2)^2}{2} + i x_5 (T - x_2) + i x_6\right), \tag{5.12}$$

where x_1, x_2, x_3, x_4, x_5, and x_6 represent the pulse amplitude, temporal position, pulse width, chirp, frequency shift, and phase, respectively. Using the generalized POM on equation (5.8), the pulse parameter equations of motion corresponding to the bare approximation of the CV theory can be obtained. The pulse parameter equation (5.13) is useful in investigations of hyperbolic secant-shaped nonlinear pulse propagation in PCF:

$$\begin{aligned}
\frac{dx_1}{dz} &= \frac{1}{2} x_4 \beta_2(z) x_1 - 3B_3(z) x_4 x_5 x_1 + \frac{1}{10} B_4(z) x_4 \left(60 x_5^2 + \frac{\left(\pi^2 x_4^2 x_3^4 + 4\right)\epsilon_1}{x_3^2}\right) x_1 - \Gamma x_1, \\
\frac{dx_2}{dz} &= -x_5 \beta_2(z) + \frac{B_3(z)\left(x_4^2 \epsilon_2 x_3^4 + 60 x_5^2 x_3^2 + 28\right)}{20 x_3^2} - \frac{B_4(z) x_5 \left(x_4^2 \epsilon_2 x_3^4 + 20 x_5^2 x_3^2 + 28\right)}{5 x_3^2} \\
&\quad + \frac{6}{5} \gamma_2 x_1^2, \\
\frac{dx_3}{dz} &= -x_3 x_4 \beta_2(z) + 6B_3(z) x_3 x_4 x_5 + B_4(z)\left(-12 x_3 x_4 x_5^2 - \frac{x_4\left(\pi^2 x_4^2 x_3^4 + 4\right)\epsilon_1}{5 x_3}\right), \\
\frac{dx_4}{dz} &= \beta_2(z)\left(x_4^2 - \frac{\epsilon_3}{x_3^4}\right) + 6B_3(z) x_5 \left(\frac{\epsilon_3}{x_3^4} - x_4^2\right) + \\
&\quad \frac{B_4(z)\left(13\pi^6 x_4^4 x_3^8 + x_4^2 \epsilon_4 x_3^4 + 84\left(\pi^4 x_3^4 x_4^2 - 30\right) x_5^2 x_3^2 - 840\right)}{7\pi^4 x_3^6} - \frac{\gamma_1 \epsilon_3 x_1^2}{x_3^2} + \frac{x_5 \gamma_2 \epsilon_3 x_1^2}{x_3^2} \\
\frac{dx_5}{dz} &- \frac{B_3(z) x_4 \left(\pi^2 x_3^4 x_4^2 - 4\right)\epsilon_5}{2 x_3^2} + \frac{2B_4(z) x_4 \left(\pi^2 x_3^4 x_4^2 - 4\right) x_5 \epsilon_5}{x_3^2} + x_4 \gamma_2 \epsilon_6 x_1^2 + \frac{4\gamma_3 x_1^2}{\pi^2 x_3^2} \\
\frac{dx_6}{dz} &= \beta_2(z)\left(\frac{\epsilon_7}{x_3^2} - \frac{x_5^2}{2}\right) + \frac{B_3(z) x_5 \left(5\pi^2\left(x_3^2 \epsilon_{11} x_4^2 + 8 x_5^2\right) x_3^2 + \epsilon_{10}\right)}{20\pi^2 x_3^2} + \gamma_1 \epsilon_8 x_1^2 + x_5 \gamma_2 \epsilon_9 x_1^2 + \\
&\quad \frac{B_4(z)\left(3\left(-27\pi^6 x_4^4 x_3^6 - x_4^2 x_5^2 \epsilon_{14} x_3^4 + 840\pi^2\left(x_4^2 - 2x_5^4\right) x_3^2 + x_5^2 \epsilon_{13}\right) x_3^2 + \epsilon_{12}\right)}{1680\pi^2 x_3^4},
\end{aligned} \tag{5.13}$$

where

$$\begin{aligned}
\Gamma &= \alpha/2,\ B_3(z) = \beta_3(z)/6,\ B_4(z) = \beta_4(z)/24,\ \gamma_1 = \gamma,\ \gamma_2 = \gamma/\omega_0,\ \gamma_3 = \gamma T_R, \\
\epsilon_1 &= \frac{-15 + 7\pi^2}{3 + \pi^2},\ \epsilon_2 = 5(-6 + \pi^2),\ \epsilon_3 = \frac{30}{\pi^4}, \\
\epsilon_4 &= 28\pi^2(-15 + \pi^2),\ \epsilon_5 = \frac{15 + \pi^2}{5\pi^2},\ \epsilon_6 = \frac{4(15 + 2\pi^2)}{15\pi^2},\ \epsilon_7 = \frac{1}{6} + \frac{5}{4\pi^2}, \\
\epsilon_8 &= \frac{15 + 8\pi^2}{12\pi^2},\ \epsilon_9 = \frac{-75 + 32\pi^2}{60\pi^2},\ \epsilon_{10} = -150 + 8\pi^2,\ \epsilon_{11} = -6 + \pi^2, \\
\epsilon_{12} &= 112(75 + 7\pi^2),\ \epsilon_{13} = 336(25 - 6\pi^2),\ \epsilon_{14} = 560\pi^2(-6 + \pi^2).
\end{aligned} \tag{5.14}$$

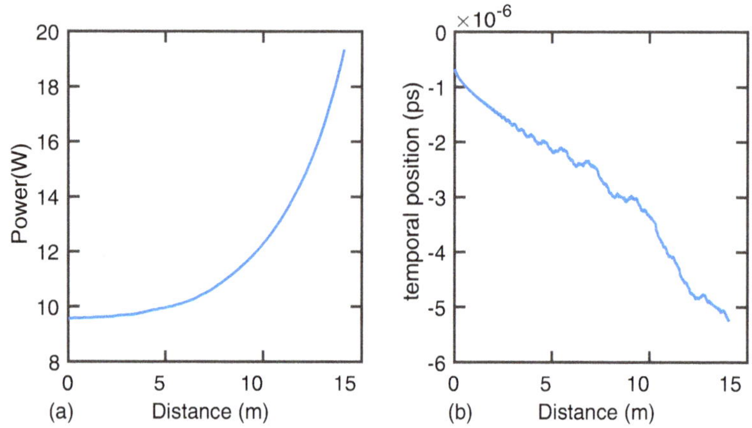

Figure 5.10. Variation in the dynamics of pulse parameters calculated using the POM along the fiber length showing (a) power and (b) temporal position.

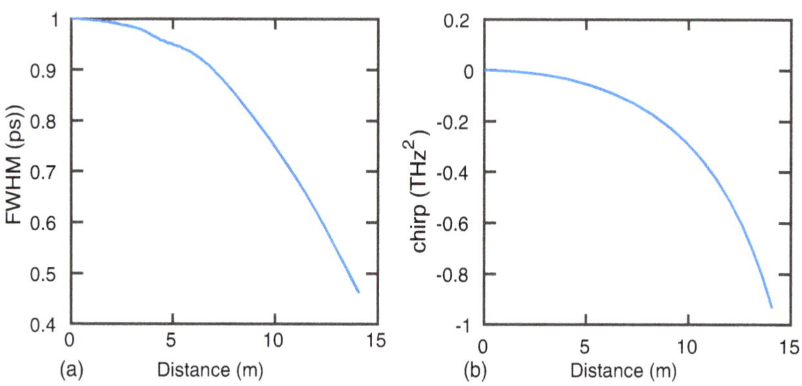

Figure 5.11. Variation in the dynamics of pulse parameters calculated using the POM along the fiber length showing (a) pulse width and (b) chirp.

The above set of ordinary differential equations gives us an idea of how the different pulse parameters are related to the fiber parameters. As a result, we can obtain greater insight into the dynamics of pulse propagation. The evolution of the pulse parameters along the propagation distance 'z' can be determined by solving equation (5.13) using the Runge-Kutta method. The dynamics of the pulse parameters during pulse propagation are shown in figures 5.10(a) and (b), 5.11(a) and (b), 5.12(a) and (b) and 5.13(a) and (b) for power, temporal position, width, chirp, frequency phase, and linear and nonlinear lengths, respectively.

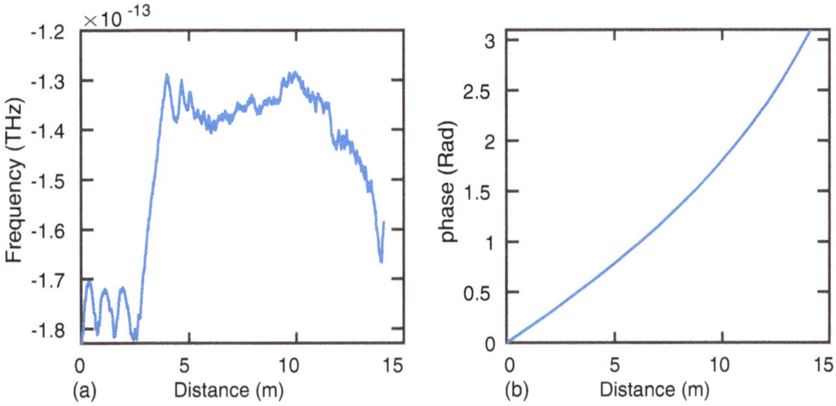

Figure 5.12. Variation in the dynamics of pulse parameters calculated using the POM along the fiber length showing (a) frequency and (b) phase.

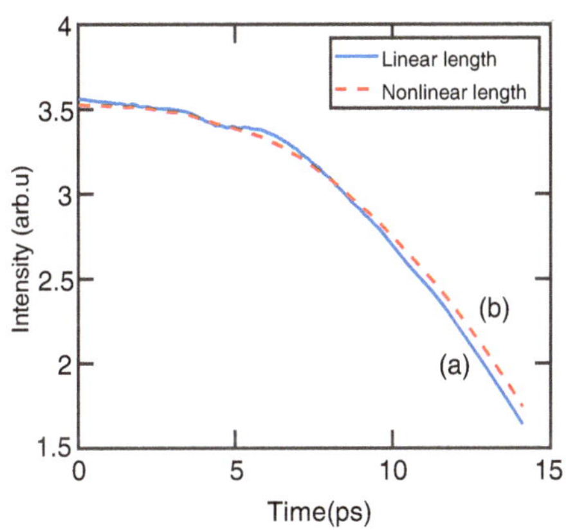

Figure 5.13. Variation in the dynamics of pulse parameters calculated using the POM along the fiber length showing (a) linear and (b) nonlinear lengths.

References

[1] Inoue T and Namiki S 2008 Pulse compression techniques using highly nonlinear fibers *Laser Photon. Rev.* **2** 83–99

[2] Agrawal G 2001 *Applications of Nonlinear Fiber Optics* 1 (Amsterdam: Elsevier) https://www.elsevier.com/books/applications-of-nonlinear-fiber-optics/agrawal/978-0-12-045144-9

[3] Ouzounov D G, Hensley C J, Gaeta A L, Venkateraman N, Gallagher M T and Koch K W 2005 Soliton pulse compression in photonic band-gap fibers *Opt. Express* **13** 6153–9

[4] Fu L, Fuerbach A, Littler I C M and Eggleton B J 2006 Efficient optical pulse compression using chalcogenide single-mode fibers *Appl. Phys. Lett.* **88** 081116

[5] Vinoj M N and Kuriakose V C 2003 Generation of pedestal-free ultrashort soliton pulses and optimum dispersion profile in real dispersion-decreasing fibre *J. Opt. A: Pure Appl. Opt.* **6** 63

[6] Voronin A A and Zheltikov A M 2008 Soliton-number analysis of soliton-effect pulse compression to single-cycle pulse widths *Phys. Rev.* A **78** 063834
[7] Cao W-h and Wai P K A 2005 Picosecond soliton transmission by use of concatenated gain-distributed nonlinear amplifying fiber loop mirrors *Appl. Opt.* **44** 7611–20
[8] Agrawal G P 2006 *Nonlinear Fiber Optics* 4 (Cambridge, MA: Academic Press)
[9] Dixit N and Vijaya R 2004 The role of positive pre-chirping on the nonlinear compression of a dual frequency beat signal with a dispersion imbalanced loop mirror *J. Opt. A: Pure Appl. Opt.* **6** 412
[10] Wai P K A and Cao W-h 2003 Ultrashort soliton generation through higher-order soliton compression in a nonlinear optical loop mirror constructed from dispersion-decreasing fiber *J. Opt. Soc. Am.* B **20** 1346–55
[11] González-Baquedano N, Torres-Gómez I, Arzate N, Ferrando A and Ceballos-Herrera D E 2013 Pulse quality analysis on soliton pulse compression and soliton self-frequency shift in a hollow-core photonic bandgap fiber *Opt. Express* **21** 9132–43
[12] Chernikov S V, Dianov E M, Richardson D J and Payne D N 1993 Soliton pulse compression in dispersion-decreasing fiber *Opt. Lett.* **18** 476–8
[13] Smith K and Mollenauer L F 1989 Experimental observation of adiabatic compression and expansion of soliton pulses over long fiber paths *Opt. Lett.* **14** 751–3
[14] Chernikov S V and Mamyshev P V 1991 Femtosecond soliton propagation in fibers with slowly decreasing dispersion *J. Opt. Soc. Am.* B **8** 1633–41
[15] Blow K J and Wood D 1988 Mode-locked lasers with nonlinear external cavities *J. Opt. Soc. Am.* B **5** 629–32
[16] Tajima K 1987 Compensation of soliton broadening in nonlinear optical fibers with loss *Opt. Lett.* **12** 54–6
[17] Nakazawa M, Yoshida E, Kubota H and Kimura Y 1994 Generation of a 170 fs, 10 GHz transform-limited pulse train at 1.55 μm using a dispersion-decreasing, erbium-doped active soliton compressor *Electron. Lett.* **30** 2038–40
[18] Kuehl H H 1988 Solitons on an axially nonuniform optical fiber *J. Opt. Soc. Am.* B **5** 709–13
[19] Ming-Leung V T, Horak P, Poletti F and Richardson D J 2008 Designing tapered holey fibers for soliton compression *IEEE J. Quantum Electron.* **44** 192–8
[20] Hu J, Marks B S, Menyuk C R, Kim J, Carruthers T F, Wright B M, Taunay T F and Friebele E J 2006 Pulse compression using a tapered microstructure optical fiber *Opt. Express* **14** 4026–36
[21] Travers J C, Stone J M, Rulkov A B, Cumberland B A, George A K, Popov S V, Knight J C and Taylor J R 2007 Optical pulse compression in dispersion decreasing photonic crystal fiber *Opt. Express* **15** 13203–11
[22] Dudley J M and Taylor J R 2009 Ten years of nonlinear optics in photonic crystal fibre *Nat. Photon.* **3** 85–90
[23] Saitoh K and Koshiba M 2002 Full-vectorial imaginary-distance beam propagation method based on a finite element scheme: application to photonic crystal fibers *IEEE J. Quantum Electron.* **38** 927–33
[24] COMSOL Multiphysics 1998 Introduction to comsol multiphysics *COMSOL Multiphysics, Burlington, MA* (accessed February 9, 2018)
[25] Kutz J N, Holmes P, Evangelides S G and Gordon J P 1998 Hamiltonian dynamics of dispersion-managed breathers *J. Opt. Soc. Am.* B **15** 87–96

[26] Raja R V J, Porsezian K, Varshney S K and Sivabalan S 2010 Modeling photonic crystal fiber for efficient soliton pulse propagation at 850 nm *Opt. Commun.* **283** 5000–6

[27] Nakkeeran K and Wai P K A 2007 *Behavior of different ansätze in the generalized projection operator method Chaos Solitons Fractals* **31** 639–47

[28] Nakkeeran K and Wai P K A 2005 Generalized projection operator method to derive the pulse parameters equations for the nonlinear Schrödinger equation *Opt. Commun.* **244** 377–82

[29] Moubissi A B, Nakkeeran K, Dinda P T and Kofane T C 2001 Non-Lagrangian collective variable approach for optical solitons in fibres *J. Phys. A: Math. Gen.* **34** 129

[30] Kamagate A, Grelu P H, Tchofo-Dinda P, Soto-Crespo J M and Akhmediev N 2009 Stationary and pulsating dissipative light bullets from a collective variable approach *Phys. Rev.* E **79** 026609

Chapter 6

Self-similar compression

This chapter discusses a pulse-compression scheme based on self-similar analysis that aims to achieve large compression factors and minimal pedestal energies by compressing fundamental solitons with low input pulse energy over small propagation distances. A numerical model of a photonic crystal fiber (PCFs) compressor that uses self-similar properties is discussed in detail.

6.1 Review of pulse compression

As seen in the previous chapters, the generation of ultrashort pulses that have the desired pulse widths has become quite possible through pulse-compression techniques with the advent of PCF. Although soliton compression can provide a large degree of compression, the compressed pulses suffer from significant pedestal generation, leading to nonlinear interactions between neighboring solitons. Needless to say, pedestal-free ultrashort pulses are desirable for typical uses in telecommunications and sensing. In this regard, the compression of fundamental solitons with no pedestal component has been achieved by the alternative adiabatic compression technique using DDFs obtained by tapering [1–3]. However, it is difficult to maintain the adiabatic condition to preserve the soliton during the whole compression process. Furthermore, it is still difficult to suppress the pedestal completely using this technique. These problems mainly arise due to random variations in the decreasing dispersion and increasing nonlinearity along the tapered fiber [1, 2]. Various dispersion-decreasing profiles have been considered for adiabatic compression schemes, such as linear, hyperbolic, exponential, logarithmic, and Gaussian [4–8]. It is certainly the case that if one can choose suitable dispersion and nonlinearity variations along the fiber, highly compressed pedestal-free pulses can be achieved by preserving the soliton shape in an appropriate way [9]. Hence, a more advanced theory was needed to better predict the dispersion-decreasing and nonlinearity-increasing profiles in order to achieve all-fiber clean ultrashort pedestal-free pulse compression.

doi:10.1088/978-0-7503-2686-5ch6

In this context, Moores introduced a novel pulse-compression technique using numerical simulation which was similar to adiabatic soliton amplification [10]. He pointed out that exact chirped soliton solutions to the NLS equation exist when the group velocity dispersion (GVD) parameters have exponentially varying dispersion $\beta_2(z) = \beta_{20}\exp(-z/z_0)$. The parameters β_2 and β_{20} are the second-order dispersion (SOD) coefficients at the distances z and $z = 0$. One of the advantages of this compression scheme is that the adiabatic condition does not need to be satisfied and rapid compression is possible. Using this approach, the pulses exhibit no radiative loss, conserve their areas, and maintain a constant chirp relative to the pulse width. More recently, a technique known as self-similar analysis has been utilized to study linearly chirped pulses in optical fibers and fiber Bragg gratings [11–13].

Self-similar pulses are preferred because of the presence of linear chirp, which facilitates efficient pulse compression. The soliton dynamics in self-similarly modeled fiber that has longitudinally varying dispersion parameters with constant gain revealed that the interplay of dispersion, nonlinearity, and gain produces a linearly chirped pulse which resists the deleterious effects of optical wave breaking. Thus, the wider study of self-similar dynamics in fibers with longitudinally varying parameters is relevant to understand pulse shaping and pulse-compression applications [14, 15]. Hence, this chapter is intended to provide a detailed description of self-similar pulse compression.

6.2 Pulse compression through self-similar analysis

6.2.1 Why use self-similar scale analysis in pulse compression?

Scale invariance at any magnification level denotes a self-similar property. In the self-similarity approach, an object that has micro dimensions maintains its overall shape even after a large magnification has been applied, and only exhibits a change in the length scale. To put this in a better way, in the self-similarity approach, the length scale varies at different positions/times, but the shape remains same. Despite its apparent complexity, self-similarity in simple terms relies on the concept of 'similar triangles,' which possess the same angles but have sides of different lengths. In such a case, it is straightforward to map one triangle on to another by a simple linear scaling transformation. A self-similar solution is one in which the functional form of the solution is invariant, so that the solution at any given point can be found from knowledge of the solution at another point by a similarity transformation. Figure 6.1 depicts a simple example of the self-similar concept in tree branches; here, each branch appears similar even at a wide range of magnification scales. In the generation of ultrashort pulses, the main technical challenge caused by pedestal energy is the maintenance of the pulse shape during the compression process. As a result, numerous investigations have been carried out to avoid these deleterious effects in pulse compression. In order to minimize the pedestal energy in the pulse-compression process, self-similar analysis has been utilized to compress the pulse while preserving the shape by harnessing profound advancements in PCF technology. The self-similar compression process effectively reduces the pulse width by keeping the pulse shape unchanged, as per the scaling transformation.

Figure 6.1. Leaves that demonstrate the concept of self-similarity.

6.2.2 Self-similar analysis

Self-similar scaling analysis helps to enhance the compression factor and quality of compressed pulses [11, 12, 15]. In what follows, we elaborate the search for a self-similar solution for the complex field function $U(z, T)$ that has the functional form given as

$$U = \psi(z, T)\exp[i\phi(z, T)], \tag{6.1}$$

where ψ and ϕ are real functions of z and T. By substituting ansatz (6.1) into equation (5.6) and separating the real and imaginary parts, a system of two equations for the phase function ϕ and the amplitude function ψ can be written, as follows:

$$\psi\phi_z = \frac{\beta_2(z)}{2}\left[\psi\left(\frac{\partial\phi}{\partial T}\right)^2 - \frac{\partial^2\psi}{\partial T^2}\right] \tag{6.2}$$

$$\psi_z = \frac{\beta_2(z)}{2}\left[\psi\frac{\partial^2\phi}{\partial T^2} + 2\frac{\partial\psi}{\partial T}\frac{\partial\phi}{\partial T}\right]. \tag{6.3}$$

Let us assume that

$$\phi(z, T) = \alpha_1(z) + \alpha_2(z)(T - T_c)^2. \tag{6.4}$$

Applying equation (6.4) in equation (6.2), we get

$$\psi\left(\frac{\partial \alpha_1}{\partial z} + \frac{\partial \alpha_2}{\partial z}(T - T_c)^2\right) = 2\beta_2(z)\psi\alpha_2^2(T - T_c)^2 - \frac{\beta_2(z)}{2}\frac{\partial^2\psi}{\partial T^2} + \gamma\psi^3. \tag{6.5}$$

Here, equation (6.5) contains an explicit dependence on the variable value $(T - T_c)$, which disappears when the terms of the monomial $(T - T_c)^2$ are equal, hence we obtain the pair of equations. By comparing the coefficients of $(T - T_c)^2$, we obtain

$$\frac{\partial \alpha_2(z)}{\partial z} = 2\beta_2(z)\alpha_2^2(z), \tag{6.6}$$

and by comparing the constants, we obtain

$$\psi\frac{\partial \alpha_1(z)}{\partial z} = -\frac{\beta_2(z)}{2}\frac{\partial^2\psi}{\partial T^2} + \gamma(z)\psi^3. \tag{6.7}$$

Similarly, applying equation (6.4) in equation (6.3) yields

$$\frac{\partial \psi}{\partial z} = \beta_2(z)\alpha_2(z)\psi + 2\beta_2(z)\alpha_2(z)(T - T_c)\frac{\partial \psi}{\partial T}. \tag{6.8}$$

To proceed further, let us consider a 'self-similar' ansatz such as

$$\psi(z, T) = \frac{1}{\sqrt{\Gamma(z)}}R(\tau)\exp\left(\frac{G(z)}{2}\right), \tag{6.9}$$

where we are aimed to determine the functions $\Gamma(z)$, $G(z)$ and $R(\tau)$, and the scaling variable τ and the function $G(z)$ are

$$\begin{aligned} \tau &= \frac{T - T_c}{\sqrt{\Gamma(z)}} \\ G(z) &= \int_0^z g(z')dz', \end{aligned} \tag{6.10}$$

where, without loss of generality, we can assume that $\Gamma(0) = 1$. By substituting equation (6.9) into equation (6.8) and comparing the coefficient of $R(\tau)$, we then get

$$\frac{1}{\Gamma(z)}\frac{\partial \Gamma(z)}{\partial z} = -2\beta_2(z)\alpha_2(z). \tag{6.11}$$

The solutions of equation (6.6) and equation (6.11) are found to be

$$\alpha_2(z) = \frac{\alpha_{20}}{1 - \alpha_{20}D(z)} \tag{6.12}$$

and

$$\Gamma(z) = 1 - C_0 D(z), \tag{6.13}$$

where

$$D(z) = 2\int_0^z \beta_2(z')dz'. \tag{6.14}$$

Similarly, by substituting equation (6.9) into equation (6.7), we get

$$\frac{d^2R}{d\tau^2} + 2\frac{2\Gamma^2(z)}{\beta_2(z)}\frac{d\alpha_1}{dz}R - \frac{2\Gamma(z)\gamma(z)}{\beta_2(z)}\exp(G(z))R^3. \tag{6.15}$$

Equation (6.15) has a non-trivial solution ($R(\tau) \neq 0$) if and only if the coefficients in equation (6.15) are constants, i.e.

$$-\frac{2\Gamma^{(}z)}{\beta_2(z)}\frac{d\alpha_1}{dz} = \lambda_1 \tag{6.16}$$

and

$$-\frac{2\gamma(z)\Gamma(z)}{\beta_2(z)}\exp(G(z)) = \alpha. \tag{6.17}$$

Here, λ_1 and α are constants. Hence, equations (6.16) and (6.17) yield

$$\begin{aligned} \lambda_1 &= \frac{-2}{\beta_2(0)}\frac{d\alpha_1}{dz}\Big|_{z=0} \\ \alpha &= \frac{\gamma(0)}{\beta_2(0)}, \end{aligned} \tag{6.18}$$

because $\Gamma(0) = 1$ and $G(0) = 0$. Therefore, in the non-trivial case, equation (6.15) can be written as

$$\frac{d^2F}{d\tau^2} - \lambda_1 F - 2\alpha F^3 = 0. \tag{6.19}$$

By integrating and equating the solution of equation (6.16), we get

$$\alpha_1(z) = \alpha_{10} - \frac{\lambda_1}{2}\int_0^z \frac{\beta_2(z')}{1 - \alpha_{20}D(z')}dz'. \tag{6.20}$$

Here, α_{10} is constant. Using relation 6.14, equation (6.20) can be written as

$$\alpha_1(z) = \alpha_{10} - \frac{\lambda_1}{4}\frac{D(z)}{1 - C_0 D(z)}. \tag{6.21}$$

Substituting all the values in equation (6.4) yields

$$\phi(z, T) = \alpha_{10} - \frac{\lambda_1}{4}\frac{D(z)}{[1 - C_0 D(z)]} + \frac{C_0(T - T_c)^2}{[1 - C_0 D(z)]}. \tag{6.22}$$

Multiplying equation (6.19) by $\frac{dF}{d\tau}$, the following equation can be obtained:

$$\frac{dF}{d\tau}\frac{d^2F}{d\tau^2} - \lambda_1 F\frac{dF}{d\tau} - 2\alpha F^3\frac{dF}{d\tau} = 0. \tag{6.23}$$

equation (6.23) can be simplified to

$$\left(\frac{dF}{d\tau}\right)^2 = \alpha F^4 + \lambda_1 F^2 + C. \tag{6.24}$$

For a bright soliton such that $C = 0$, $\alpha < 0$,

$$F = \sqrt{\frac{\lambda_1}{\alpha}}\,\mathrm{sech}(\sqrt{\lambda_1}\,\tau). \tag{6.25}$$

As we know, $\lambda_1 = \frac{1}{T_0^2}$ and $\alpha = \frac{1 - c_0 D(z)\exp(G(z))}{\rho(z)}$ where $\rho(z) = \frac{\beta_2(z)}{\gamma(z)}$, ψ takes the form

$$\psi(z, T) = \frac{\sqrt{\rho(z)}}{T_0[1 - C_0 D(z)]}\,\mathrm{sech}\left[\frac{T - T_c}{T_0[1 - \alpha_{20} D(z)]}\right]. \tag{6.26}$$

The self-similar solution is possible if and only if the following conditions are satisfied [11]:

$$\beta_2(z) = \beta_{20}\exp(-\sigma z) \tag{6.27}$$

$$\sigma = \alpha_{20}\beta_{20} \tag{6.28}$$

$$\gamma(z) = \gamma_0\exp(\rho z), \tag{6.29}$$

where β_{20} and γ_0 are the initial dispersion and nonlinearity, and ρ is the growth rate. The envelope U can then be written as

$$U(z, T) = \sqrt{\frac{|\beta_2(z)|}{\gamma(z)}}\frac{1}{T_0[1 - \alpha_{20} D(z)]} \times \mathrm{sech}\left[\frac{T - T_c}{T_0[1 - \alpha_{20} D(z)]}\right] \exp\left[i\alpha_1(z) - \frac{i\lambda_1 D(z)}{4[1 - \alpha_{20} D(z)]} + i\frac{\alpha_2(z)}{2}(T - T_c)^2\right]. \tag{6.30}$$

6.2.3 Designing PCF using self-similar analysis

As per the analytical discussion, the proposed compression scheme is described in figure 6.2, and is known as the *self-similarity-based pulse-compression technique*. In order to design new PCFs based on the analytical results, equations (6.27) and (6.29) are mainly utilized, which state that efficient pedestal-free soliton pulse compression is possible using PCFs with exponentially decreasing dispersion and exponentially increasing nonlinearity profiles. The required dispersion-decreasing and

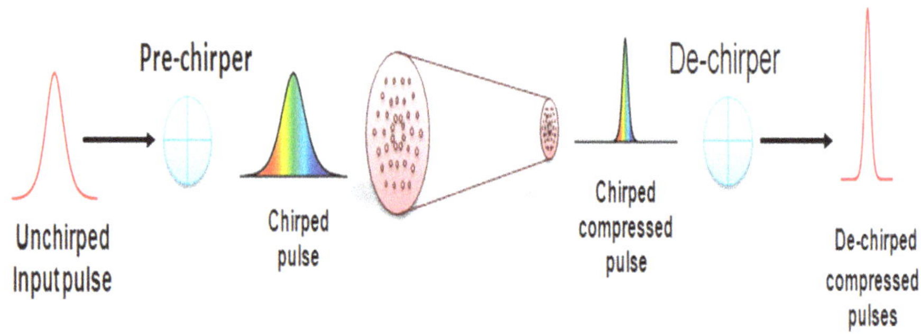

Figure 6.2. Schematic diagram of a PCF compressor based on the self-similar technique.

nonlinearity-increasing PCFs can be obtained by exponentially varying the air-hole size and pitch, as shown in figures 6.3(a) and (b). In order to accomplish self-similar compression, the physical parameter d is varied exponentially from 0.91 to 0.73, and another physical parameter Λ is varied from 1.27 μm to 1.03 μm. When choosing PCF parameters for the design of a tapered PCF, one has to make sure that the dispersion decreases from the maximum possible value within the single-mode regime. Here, the plot of the air-hole size varies exponentially, which allows a dispersion-decreasing and nonlinearity-increasing PCF to be obtained, as shown in figures 6.3(c) and (f). It should be noted that the dispersion and nonlinearity vary exponentially as required by the self-similar analysis.

6.2.4 Pedestal-free pulse compression

We have already derived the bright chirped solitary-pulse solution in equation (6.30) using self-similar analysis for the modified NLSE given in equation (5.6). The resulting chirped solitary pulse can be used to achieve pedestal-free pulse compression in PCF at 1064 nm. In order to discuss the pulse compression, we assume a PCF that has a length $L = 4L_D$. The nonlinear coefficient of silica is 2.3×10^{-20} m^2 W^{-1}. The GVD of the PCF is calculated using the FVEIM; its value is 157 ps/nm/km. The physical parameters for the self-similar scheme are $T_0 = 1$ ps and $\beta_2(0) = -0.0913$ ps^2 m^{-1}. The parameters for the adiabatic scheme are the same as those shown in figure 5.6. The length of the PCF is 4 L_D for both the adiabatic scheme and the self-similar scheme. The calculated linear length of the PCF is 14.09 m for a given input pulse width FWHM of 1 ps, hence the compression can be obtained using a shorter fiber length than the length of ordinary conventional fiber required. Using the FVEIM, we calculated the effective refractive index as a function of the propagation length for a pumping wavelength of 1064 nm. We changed the diameter of the air hole and pitch as depicted in figure 6.3. As the air-hole diameter and pitch are much more sensitive along the fiber, the effective refractive index also varies significantly as the air-hole diameter and pitch changes. Consequently, the GVD, the third-order dispersion coefficient β_3, and the nonlinearity of the proposed PCF vary with distance and have been calculated from the effective refractive indices. From figure 6.3, it can be seen that β_2 can be tuned from $-0.090\,98$ ps^2 m^{-1} to

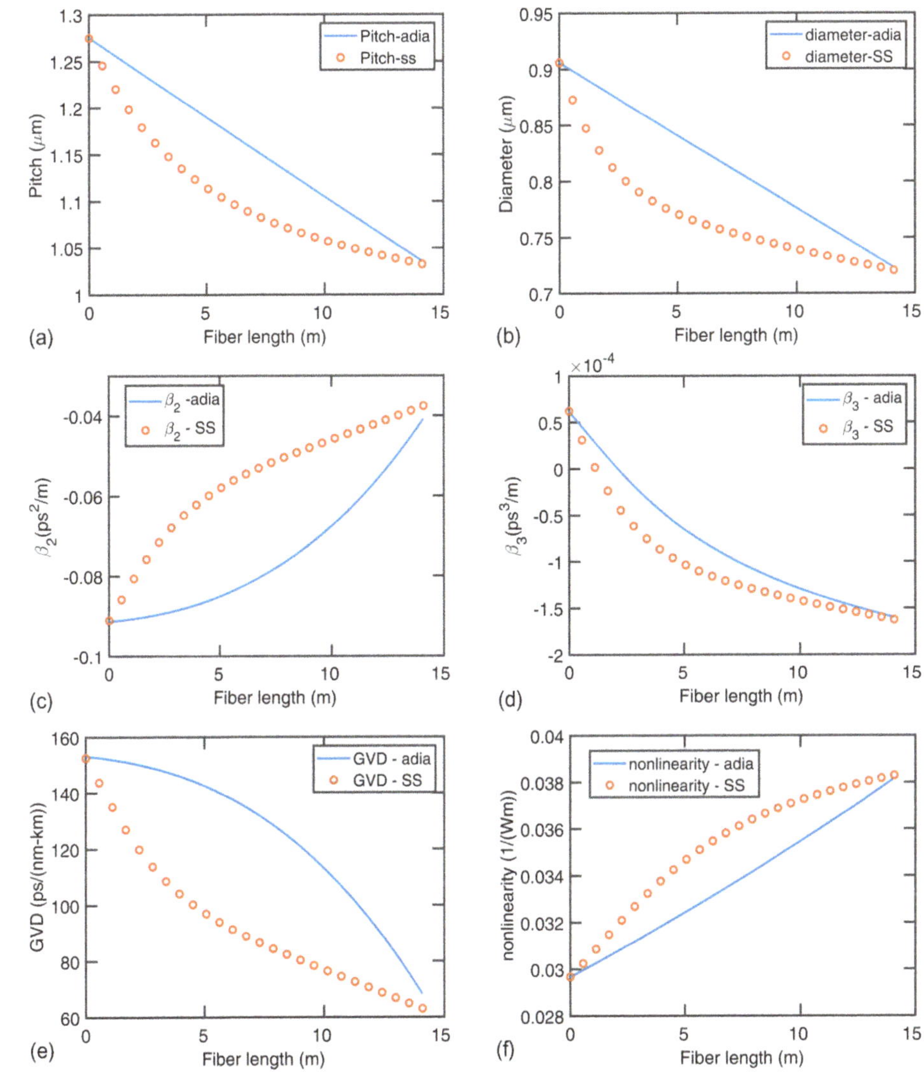

Figure 6.3. Variation of PCF parameters along the propagation distance for the adiabatic and self-similar schemes: (a) Λ, (b) d, (c) β_2, (d) β_3, (e) GVD, and (f) nonlinearity of PCF.

−0.037 58 $ps^2\ m^{-1}$ and that β_3 can be tuned from $6.22 \times 10^{-5}\ ps^3\ m^{-1}$ to $-1.62 \times 10^{-4}\ ps^3\ m^{-1}$ in the anomalous dispersion regime. Similarly, the nonlinearity can be tuned from 0.0296 $(Wm)^{-1}$ to 0.0383 $(Wm)^{-1}$.

From the dispersion and nonlinear data obtained using the FVEIM, the decay rate σ (= 0.059 94 m^{-1}) and the growth rate ρ were calculated by means of the curve-fitting method. Using equation (6.28), the chirp value was calculated to be $\alpha_{20} = \sigma/\beta_2(0)$. The evolutions of the compressed pulses that occur in the self-similar analysis and adiabatic schemes are shown in figure 6.4(a). It can clearly be observed

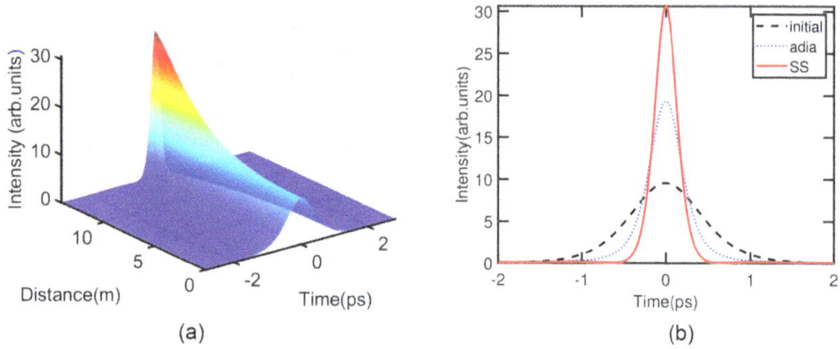

Figure 6.4. (a) Evolution of a fundamental soliton along the self-similarly modeled fiber length, (b) calculated output pulse profile for the self-similar scheme in PCF at 1064 nm compared with the adiabatic pulse compression at 1064 nm discussed in the previous chapter.

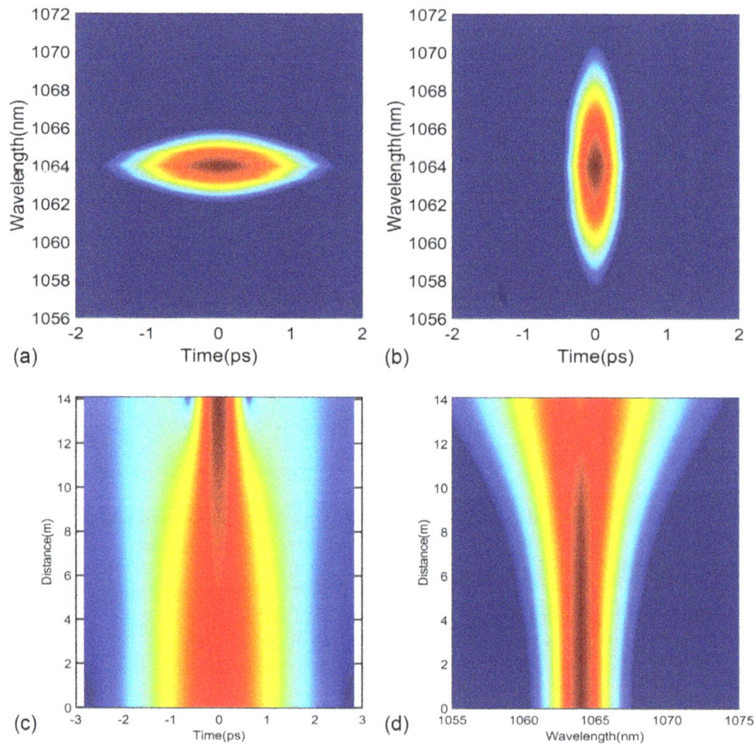

Figure 6.5. Comparison of the spectrogram profiles of (a) a 1 ps pulse and (b) a 0.31 ps pulse; schematic representation of the (c) temporal evolution and (d) spectral evolution of a propagating fundamental soliton.

from figure 6.4(b) that one can obtain a highly compressed pulse by using linearly varying chirp through self-similar analysis (figure 6.5). Figure 6.6(a) allows a comparison to be made between the compression factors and pedestal energies achieved by the adiabatic and self-similar schemes. It has been calculated that the

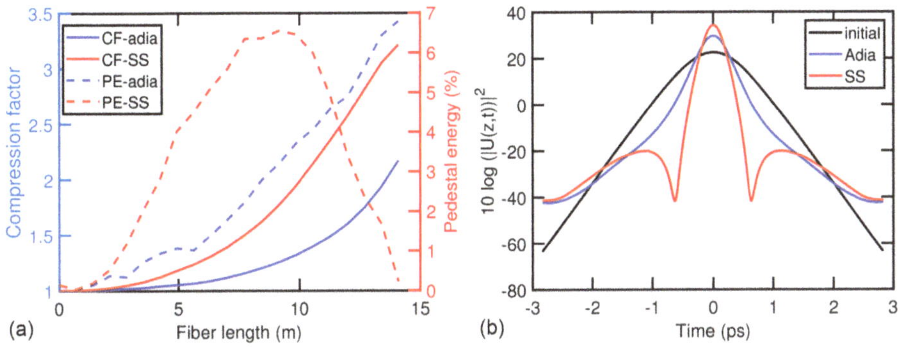

Figure 6.6. (a) Variation in the compression factor and pedestal energy obtained using self-similar and adiabatic compression schemes, (b) intensities of the compressed pulses produced using the self-similar and adiabatic schemes on a logarithmic scale.

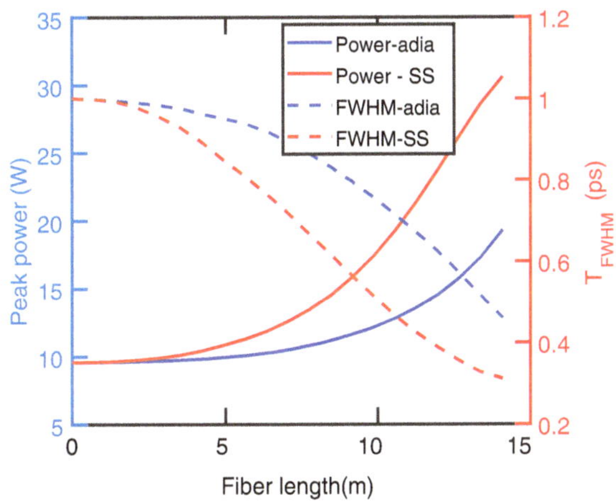

Figure 6.7. Evolution of the bright solitary-pulse peak power and T_{FWHM} for the physical parameters $T_0 = 1$ ps, $\beta_2(0) = -0.091\,33$ ps^2 m^{-1}.

compression factor achieved using self-similar analysis is 3.2, which is greater than that achieved using adiabatic pulse compression. It is necessary to examine whether the resulting compressed pulse is free from a pedestal. From figure 6.6(b), it is very clear that the compressed pulses obtained using self-similar analysis are almost pedestal free compared to those obtained via the adiabatic scheme. In general, the pedestal of the compressed pulse can clearly be observed on a logarithmic scale, as shown in figure 6.6(b). From figures 6.6(a) and (b), one can infer that a compressed pulse without any pedestal can be achieved by means of self-similar analysis. The evolution of the power and pulse width along the propagation distance 'z' can be ascertained by solving equation (4.36). The dynamics of the pulse-compression peak power and width are shown in figure 6.7 for structures that have different designs corresponding to self-similar analysis and the adiabatic scheme. The time–bandwidth product (TBP) of the compressed pulse is found to be 0.315, which is almost equal to the TBP of the hyperbolic secant pulse. The quality of the compressed pulse obtained using the self-similar technique is readily apparent, as the pedestal energy calculation shows that its pedestal energy is less than that obtained via adiabatic compression. The self-similar technique further ensures that the compressed pulse is free from chirp. Thus, the results allow us to conclude that the compressed pulse is of good quality, and that a de-chirper is not required in the proposed self-similar compression scheme, which is considered to be one of its major advantages.

In summary, in this chapter, to overcome the drawbacks of adiabatic compression, we have proposed a compact compressor for efficient chirp- and pedestal-free compression using a novel pulse-compression technique based on self-similar analysis. We have outlined the use of an appropriate self-similar scaling analysis to generate linearly chirped solitons in a PCF at 1064 nm in order to obtain ultrashort pulses. With the advent of this novel technique, new designs for tapered PCFs can be proposed that vary the pitch and diameter of the air holes to achieve the required exponentially increasing nonlinearity and exponentially decreasing dispersion, as dictated by self-similar analysis. The self-similar technique in engineered dispersion-decreasing PCF structures offers a promising way to achieve ultrashort pulses with no pedestal (or less pedestal than in conventional schemes). As a result of this new self-similar technique, efficient pulse compression can be achieved at 1064 nm with low input pulse energy over short propagation distances with a large compression factor and minimal pedestal energy compared to those achieved by the adiabatic compression scheme.

References

[1] Hu J, Marks B S, Menyuk C R, Kim J, Carruthers T F, Wright B M, Taunay T F and Friebele E J 2006 Pulse compression using a tapered microstructure optical fiber *Opt. Express* **14** 4026–36

[2] Travers J C, Stone J M, Rulkov A B, Cumberland B A, George A K, Popov S V, Knight J C and Taylor J R 2007 Optical pulse compression in dispersion decreasing photonic crystal fiber *Opt. Express* **15** 13203–11

[3] Nguyen H C, Kuhlmey B T, Mägi E C, Steel M J, Domachuk P, Smith C L and Eggleton B J 2005 Tapered photonic crystal fibres: properties, characterisation and applications *Appl. Phys.* B **81** 377–87

[4] Dudley J M and Taylor J R 2009 Ten years of nonlinear optics in photonic crystal fibre *Nat. Photon.* **3** 85–90

[5] Inoue T and Namiki S 2008 Pulse compression techniques using highly nonlinear fibers *Laser Photon. Rev.* **2** 83–99

[6] Ming-Leung V T, Horak P, Poletti F and Richardson D J 2008 Designing tapered holey fibers for soliton compression *IEEE J. Quantum Electron.* **44** 192–8

[7] Schenkel B, Paschotta R and Keller U 2005 Pulse compression with supercontinuum generation in microstructure fibers *J. Opt. Soc. Am.* B **22** 687–93

[8] Türke D, Wohlleben W, Teipel J, Motzkus M, Kibler B, Dudley J and Giessen H 2006 Chirp-controlled soliton fission in tapered optical fibers *Appl. Phys.* B **83** 37–42

[9] Knight J C, Birks T A, Russell P St J and Atkin D M 1996 All-silica single-mode optical fiber with photonic crystal cladding *Opt. Lett.* **21** 1547–9
[10] Moores J D 1996 Nonlinear compression of chirped solitary waves with and without phase modulation *Opt. Lett.* **21** 555–7
[11] Kruglov V I, Peacock A C and Harvey J D 2005 Exact solutions of the generalized nonlinear Schrödinger equation with distributed coefficients *Physi. Rev.* E **71** 056619
[12] Li Q, Senthilnathan K, Nakkeeran K and Wai P K A 2009 Nearly chirp- and pedestal-free pulse compression in nonlinear fiber Bragg gratings *J. Opt. Soc. Am.* B **26** 432–43
[13] Senthilnathan K, Li Q, Nakkeeran K and Wai P K A 2008 Robust pedestal-free pulse compression in cubic-quintic nonlinear media *Phys. Rev.* A **78** 033835
[14] Dudley J M, Finot C, Richardson D J and Millot G 2007 Self-similarity in ultrafast nonlinear optics *Nat. Phys.* **3** 597–603
[15] Wu L, Zhang J-F, Li L, Tian Q and Porsezian K 2008 Similaritons in nonlinear optical systems *Opt. Express* **16** 6352–60

Chapter 7

Pulse compression in nonlinear optical loop mirrors

This chapter briefly reviews the numerical simulation of simultaneous pulse compression and pedestal suppression in a tapered photonic-crystal-fiber (PCF)-based nonlinear optical loop mirror (NOLM) to model a highly efficient optical pulse compressor that generates femtosecond pulses.

7.1 Introduction

Pedestal formation is a major drawback in the aforementioned pulse-compression schemes that use PCF and other solid-state media. In order to suppress the pedestal, several techniques have been successfully implemented using optical fiber. Among these, the intensity discrimination technique and the nonlinear optical loop mirror (NOLM) are efficient techniques for suppression of the pedestal. In the intensity discrimination technique, the fiber nonlinear birefringence induced by a given input optical pulse is used to modify the shape of the same optical pulse. Hence, the tails of the highly intense pulse are blocked, while peak of the pulse is allowed to exit from an optical medium. NOLMs are fiber-based devices. In most published reports, NOLMs have been effectively used as fast saturable absorbers inside the cavity. All-fiber laser systems can be built using NOLMs, since they have the potential to improve mode-locking quality in laser systems. Although NOLMs have been studied for many years, their unknown properties are still being discovered. One such property is pedestal suppression. In NOLM systems, symmetry breaking leads to a phase difference between counter-propagating pulses. Recombination of such counter-propagating pulses allows the transmission of higher-intensity pulses, while the low-intensity pulses are reflected back into the loop, resulting in a pedestal-free transmitted pulse. The ability of the intensity discrimination technique and NOLMs to suppress the pedestal in the femtosecond region is limited by higher-order effects such as Raman self-scattering (RSS) and third-order dispersion (TOD). This chapter

doi:10.1088/978-0-7503-2686-5ch7

discusses soliton propagation in the tapered PCF-based NOLM configuration, which is used to suppress the pedestal and higher-order effects.

7.2 Nonlinear optical loop mirrors

The fiber loop mirror is a nonlinear optical waveguide that exploits the Kerr effect to switch the parameters of optical pulses. The NOLM was first proposed by Doran and Wood in 1988 [1] and experimentally demonstrated by Blow *et al* [2], and Islam *et al* [3]. The NOLM is especially attractive because of its relatively simple configuration and size; it can be further shortened using highly nonlinear fibers. This is the important feature of the device. In general, an NOLM device consists of a two-by-two directional fiber coupler, the outputs of which are connected to a loop of fiber. A schematic of an NOLM is shown in figure 7.1. The main advantage of this device is that it does not require interferometric alignment and has a simple state-of-the-art setup. In an NOLM made of a fiber that has uniform dispersion (β_2) and a nonlinear parameter (γ), when an optical signal enters through the input port, the single input is divided into two counter-propagating fields depending on the power coupling ratio α: $1 - \alpha$ at the fiber coupler. Here, both propagating fields encounter the same optical path length, since they follow the same path but in opposite directions. Due to fiber dispersion and nonlinear effects, the counter-propagating fields experience temporal broadening or compression, depending on the value of the group velocity dispersion (GVD) coefficient and spectral broadening due to self-phase modulation (SPM), which leads to a phase difference between the counter-propagating fields as they propagate through the loop. As a result of symmetry, the phase shift experienced by the counter-propagating pulses is same. After traveling around the loop, the counter-propagating pulses are recombined at the coupler. Since the central peak alone attains the maximum phase shift due to SPM, at the point of recombination in the coupler, high peak power transmitted outputs and low peak power reflected pulses are produced as a result of the intensity-dependent transmission characteristics of the NOLM. This interference governs both the

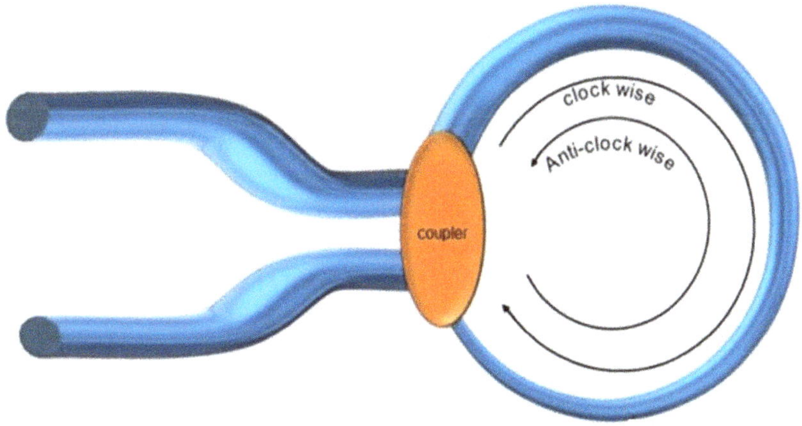

Figure 7.1. Schematic diagram of a nonlinear optical loop mirror.

optical power returned to the input port and the power exiting the output port. The interference conditions can be affected by multimode behavior, polarization shifts, and nonlinear phenomena.

By breaking the symmetry of the NOLM using symmetry-breaking elements such as attenuators, amplifiers, dispersion-varying fiber, or gain fiber, it is possible to obtain a phase difference between the propagating pulses. This makes the NOLM a self-switching device. Depending upon the configuration adopted, the NOLM can be used in various applications such as short pulse generation, optical switching, or transmission control in communications systems, saturable absorption, and pedestal suppression [4] in conventional soliton transmission systems.

7.3 Numerical model of an NOLM

The mathematical model of a sample fiber loop configuration is described in detail in this section. Here, we consider an NOLM configuration consisting of a 50:50 coupler and a dispersion-decreasing fiber (DDF) in which the dispersion coefficient decreases in the clockwise direction. The equations connecting the input and output field amplitudes in the X coupler are

$$U_c(0,\, t) = \sqrt{\alpha}\, U_{\text{in}}(0,\, t) \tag{7.1}$$

$$U_{cc}(0,\, t) = i\sqrt{(1-\alpha)}\, U_{\text{in}}(0,\, t). \tag{7.2}$$

The phase shift acquired by the counter-propagating fields in a fiber of length L under the influence of SPM is given by

$$\phi_{c,\, cc} = \frac{2\pi n_2 |U_{c,\, cc}|^2 L}{\lambda}. \tag{7.3}$$

While traveling around the fiber loop of length L, at any propagation distance z, the fields U_c and U_{cc} are given by

$$U_c(z,\, t) = \sqrt{\alpha}\, U_{\text{in}} \exp(i\alpha |U_{\text{in}}|^2 2\pi n_2 z/\lambda) \tag{7.4}$$

$$U_{cc}(z,\, t) = i\sqrt{(1-\alpha)}\, U_{\text{in}} \exp(i(1-\alpha)|U_{\text{in}}|^2 2\pi n_2 z/\lambda). \tag{7.5}$$

After propagation through the loop and recombination at the coupler, the amplitudes of the transmitted and reflected pulses are:

$$U_t(z,\, t) = \sqrt{\alpha}\, U_c(z,\, t) + i\sqrt{(1-\alpha)}\, U_{cc}(z,\, t) \tag{7.6}$$

$$U_r(z,\, t) = i\sqrt{(1-\alpha)}\, U_c(z,\, t) + \sqrt{\alpha}\, U_{cc}(z,\, t). \tag{7.7}$$

The output power transmitted by the NOLM configuration is determined by:

$$\begin{aligned} P_{\text{out}} &= |U_t|^2 \\ &= |\sqrt{\alpha}\, U_c(z,\, t) + i\sqrt{(1-\alpha)}\, U_{cc}(z,\, t)|^2. \end{aligned} \tag{7.8}$$

When the NOLM is made from a fiber that has uniform dispersion and nonlinear coefficients, for any value of $\alpha \neq 0.5$, 100% of the power emerges from output port in this configuration whenever

$$\frac{2\pi n_2 L}{\lambda} = m\frac{\pi}{1 - 2\alpha}. \tag{7.9}$$

The minimum power, which occurs for the even values of m, is given by:

$$|U_{\text{out}}|^2 = |U_{\text{in}}|^2[1 - 4\alpha(1 - \alpha)]. \tag{7.10}$$

When $\alpha = 0.5$, this device works as a mirror. If we assume that the fiber maintains polarization with no loss, during propagation, all the injected power is reflected back to the input port (i.e. the NOLM functions as a perfect reflector) over a wide wavelength range. Because of symmetry, the phase shift is same for both pulses. As a result, the reflectance remains unchanged. In this case, the length of the fiber loop is irrelevant. Similarly, external influences on the fiber loop, such as temperature variations, have no effect as long as the polarization is unchanged.

In general, pulses that undergo compression have to propagate through a relatively long fiber length. Hence, in the propagation of ultrashort pulses in NOLM, interaction between the counter-propagating fields, i.e. cross-phase modulation (XPM), has been neglected due to the short temporal duration of the pulse compared to the fiber loop length. For long pulse durations, the nonlinear interactions between the counter-propagating pulses cannot be neglected. In this case, the two waves form a diffraction grating, leading to cross interaction of the fields, and the nonlinear refractive indices are different for the two opposite directions:

$$n_c = n_2(|U_c|^2 + 2|U_{cc}|^2) \tag{7.11}$$

$$n_{cc} = n_2(2|U_c|^2 + |U_{cc}|^2). \tag{7.12}$$

Simulations of NOLMs predict that when the time–bandwidth product drops below 0.315, the transmitted pulse is positively chirped, whereas when the time–bandwidth product increases above 0.315, the transmitted pulse is negatively chirped. As well as increasing the pulse width and reducing the pulse energy, higher-order effects severely degrade the pulse quality in pedestal suppression of femtosecond pulses using NOLMs.

7.4 Applications of NOLMs

7.4.1 Amplitude equalizers

In optical pulse propagation through a rare-earth-doped fiber, a high-intensity beam experiences a larger loss than that of a lower-intensity one. In order to solve this problem, an equalizing mechanism that uses an NOLM as an amplitude equalizer has been proposed, which can effectively alleviate mode competition in rare-earth-doped fibers. As a result, the balance between the gain-clamping function of the NOLM and the mode competition effect of the rare-earth-doped fiber can lead to multiwavelength oscillations at room temperature and also ensure uniform power distribution among wavelengths. This amplitude equalizer can be employed to obtain stable multiwavelength oscillations in rare-earth-doped fiber loops at room temperature with uniform power distribution [5].

7.4.2 Saturable absorbers

Because of their strongly nonlinear properties, NOLMs have been used as saturable absorbers to passively mode lock laser oscillators, reshape optical pulses, and suppress the pedestal in the pulse-compression process, i.e. to filter out low-intensity noise and dispersive waves from higher-power signals through effective saturable absorption (a decrease in attenuation with increasing optical power). Pedestal suppression avoids pulse instabilities that are caused by the interaction of such background radiation with the carrier pulse. Among the various kinds of saturable absorber, NOLMs are promising, as they only require a low-cost fiber coupler and moderate fiber segments and can be implemented in a polarization-maintaining configuration. Fiber NOLMs have an ultrafast response in the subpicosecond range, which most saturable absorbers do not have. For instance, traditional semiconductor saturable absorbers have nanosecond response times.

NOLMs have been used as saturable absorbers by unbalancing the symmetry of the system, which can be achieved using an asymmetric coupler or alternatively with a gain/loss element asymmetrically placed in the loop; it can also be achieved by employing dispersion-decreasing fibers, nonlinearity-increasing fibers, or gain fibers [6]. Other variations include the use of two fibers with different dispersive properties within the loop [7]. The propagation of ultrashort pulses with significant peak power though an unbalanced NOLM induces different phase shifts in the counter-propagating pulses, depending on the characteristics of the symmetry-breaking element. The presence of nonlinear effects enhances the pulse peak power, which in turn influences the phase velocity. The nonlinear phase shift is not constant over the temporal pulse profile, but is greater at the peak and less so in the wings. As a result, the device reacts to the phase difference between two counter-propagating fields. Hence, the interference conditions change, and the low-power background waves are reflected back, while high-power pulses switch the NOLM into transmittance. Moreover, owing to its switching characteristic, the NOLM can stabilize the pulse amplitude. A study of the performance of an NOLM as saturable absorber in a normal-dispersion fiber laser was presented in [8]. The dual function of intensity filtering and pulse amplitude control is probably the most attractive feature that distinguishes NOLMs from other saturable absorbers.

7.5 Soliton propagation in NOLMs

Although NOLMs reduce the background waves and dispersive waves which are responsible for intersymbol interference in optical fiber communications systems, they also reduce the pulse energy through saturable absorption. In optical pulse propagation through the NOLM, SPM and GVD are the two effects that severely limit the performance of non-soliton systems [9]. In order to reduce the impact of dispersion and nonlinear effects, soliton propagation has been introduced by perfectly balancing the linear and nonlinear effects with the aid of negative GVD fibers. The formation of solitons helps to conserve the energy of the optical pulse while it propagates through the fiber. Hence, to enhance the transmission properties of NOLMs, soliton propagation in the loop has been investigated; the resulting devices are the leading candidates for long-haul light-wave transmission links.

Nonlinear soliton transmission in an NOLM made from a fiber that has a negative GVD parameter shows an enhancement in pulse energy and suppress the formation of a pedestal as compared to non-soliton systems. In order to conserve the soliton nature, a significant length of fiber is required. The fiber length required decreases with the square of the pulse duration, according to the soliton requirement ($N^2 = T_0^2\gamma P_0/|\beta_2|$). When a soliton propagates through an NOLM, the features of the NOLM are altered dramatically. For example, [10] proved that an NOLM functioning in the soliton region became stable against small fluctuations in the input power. In an unbalanced NOLM, counter-propagating solitons experiences different phase shifts depending on the symmetry-breaking element used. By appropriate adjustment of the input pulse power, almost perfectly linear chirps can be obtained [11]. This study found that suitable pre-chirping of the signal resulted in better shaping characteristics and significant pedestal suppression. The residual side peaks which survived the soliton action were then substantially reduced by the NOLM.

7.6 Soliton pulse compression in NOLMs

In practical terms, in all of the ultrashort pulse-generation techniques based on solid-state and fiber media, picosecond to subpicosecond pulses can be easily obtained through mode-locking techniques. A further reduction in temporal duration can be obtained through the soliton pulse-compression techniques described in the previous chapters. Utilizing those techniques, fundamental soliton compression alone generates ultrashort low-pedestal pulses. However, an enhancement in the compression factor can be achieved through higher-order soliton compression, which provides poor-quality pulses because of the formation of an undesirable broad pedestal. The pedestal energy is defined as the relative difference between the total energy of the transmitted pulse (E_{out}) and the energy of a hyperbolic secant pulse that has the same peak power and width as that of the transmitted pulse (E_{sech}). The formation of a pedestal decreases the effective utilization of the pulse energy. This can be understood by considering the fact that under the high-power conditions, although the pulse width of the compressed pulse is very narrow, the ratio of the energy in the pedestal to the total energy of the pulse is quite large. This means that more energy remains in the pedestal of the compressed pulse. Due to the broad pedestal, the adjacent pulses overlap during propagation, causing a system performance degradation.

In order to suppress the energy that accumulates at the wings of ultrashort pulses, pedestal suppression techniques such as spectral filtering and intensity discrimination techniques are available, but these require complex arrangements of optical components; hence, they do not suppress the pedestal effectively. In the previous sections we have seen that an unbalanced NOLM configuration can provide pedestal suppression while conserving energy for the propagation of solitons. In addition, during soliton propagation through an NOLM, pulse-width reduction takes place due to the extra nonlinearity induced by the NOLM through the intensity-dependent optical Kerr effect. By utilizing these two properties of an unbalanced NOLM, simultaneous pulse compression and pedestal suppression have been realized when the optimal conditions are satisfied for the propagation of solitons.

Initially, simultaneous pulse compression and pedestal suppression using NOLMs were demonstrated in [11–15]. These studies did not incorporate soliton propagation. When compressed pulses are output by an asymmetrical NOLM, the pedestal with more energy are moved; as a result, the residual energy in the pulse and the energy transmissivity are decreased [15]. Later, the studies described in [14] and [15] examined in detail the compression of higher-order solitons in an unbalanced NOLM constructed from a uniform fiber, where, although simultaneous pulse compression and pedestal suppression were achieved, the compressed pulse shape deviated significantly from that of a soliton. In order to achieve efficient compression, the authors of [12, 16] and [13] utilized NOLMs containing an asymmetric coupler and a uniform piece of fiber or NOLMs constructed from two different fibers with a symmetric coupler. The variation in dispersion greatly enhanced the compression process, producing significant pedestal suppression, but the general optimization criteria for pulse compression were not studied. Later, using dispersion-decreasing fibers, pulse compression and pedestal suppression were successfully studied in an NOLM configuration [11, 15]. The complete process of simultaneous pulse compression and pedestal suppression using dispersion-decreasing fiber is elaborated in the following sections.

7.6.1 Demonstration of the technique

The proposed higher-order soliton compression scheme that uses a dispersion-decreasing PCF (DD-PCF) in an NOLM configuration does not require the adiabatic condition to be satisfied; hence, it can be used to compress a longer pulse using a reasonable fiber length. Furthermore, this scheme is more tolerant of initial frequency chirps and higher-order effects than the adiabatic compression technique; hence, high-quality compression of higher-order solitons is possible. In this compression scheme, ultrashort pulse generation results from both the soliton-effect compression of the counter-propagating pulses and the switching characteristics of the DD-PCF-based NOLM. Figure 7.1 shows the configuration of an unbalanced NOLM compressor that contains DD-PCF as the symmetry-breaking element and a 50/50 coupler. In this NOLM compressor, the dispersion coefficient decreases exponentially in the clockwise direction. For the compression process, the input pulse is assumed to be a hyperbolic secant pulse with a soliton order of $N = 6$, which is represented in functional form as:

$$U(0, t) = N\mathrm{sech}(t/T_0)\exp(-iCt^2/2) \tag{7.13}$$

where the soliton order N is related to the physical parameters of the fiber by

$$N^2 = \frac{\gamma_0 P_0 T_0^2}{|\beta_2(0)|}. \tag{7.14}$$

Here, γ_0 is the nonlinearity coefficient, $\beta_2(0)$ is the GVD parameter at the input end of the DD-PCF, P_0 is the peak power, and T_0 is the width of the input pulse. The DD-PCF has been modeled using the fully vectorial effective index method (FVEIM) with $\beta_2(0) = -0.1393$ ps^2 m^{-1} and $\gamma_0 = 0.1433$ W^{-1} m^{-1}. The variation in the fiber parameters such as pitch (Λ), air-hole diameter (d), $\beta_2(z)$, $\gamma(z)$, and higher-order dispersion coefficients $\beta_3(z)$ and $\beta_4(z)$ along the fiber length is depicted in figure 7.2. From figure 7.2(b), we can clearly see an exponential variation in the β_2 parameter along the fiber length with an effective compression ratio of ($\beta_{2_{\text{in}}}/\beta_{2_{\text{out}}}$) of 2.16. Since the loss of PCF is very low, it has been neglected in the compression analysis. Here, the optimal value of $L = 29$ cm has been selected for the fiber length, which is discussed in detail in the following sections. In the modeled DD-PCF, in order to allow the propagation of sixth-order solitons, the calculated $P_0 = 108$ W, and $T_0 = 1$ ps. The energy of a hyperbolic secant pulse with a peak power of P_0 and a pulse width of T_{FWHM} is given by

$$E = 2P_0 \frac{T_{\text{FWHM}}}{1.763}. \tag{7.15}$$

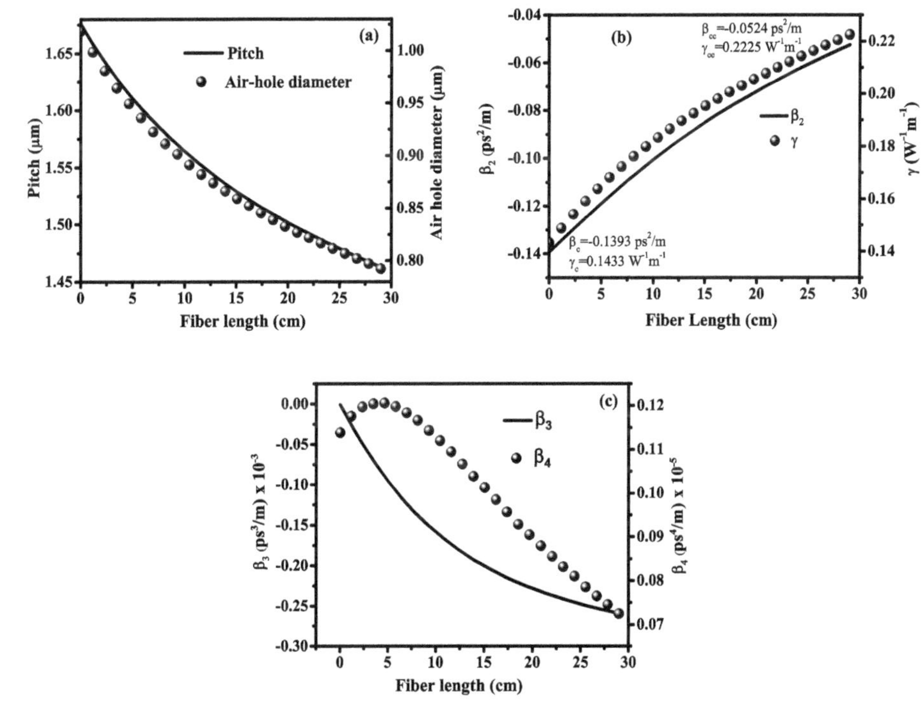

Figure 7.2. Pictorial representation of variation in the fiber parameters (a) pitch and air-hole diameter, (b) β_2 and γ, (c) β_3 and β_4.

Based on this relation, the input pulse energy is calculated to be 12 nJ. In the NOLM configuration, two counter-propagating pulses experience different dispersion profiles in the loop, i.e. the clockwise pulse initially experiences a dispersion which is smaller than that of the counterclockwise pulse. The initial values of dispersion and the nonlinear parameters for the counter-propagating pulses are $\beta_c = -0.1393$ ps^2 m^{-1}, $\gamma_c = 0.1432$ W^{-1} m^{-1} and $\beta_{cc} = -0.0524$ ps^2 m^{-1}, $\gamma_{cc} = 0.2225$ W^{-1} m^{-1}, where c denotes the clockwise-propagating pulse and cc denotes the counterclockwise-propagating pulse. The variation in pulse amplitude along the fiber length can be calculated using the relations of U_c and U_{cc}. The incidence of an input soliton of order $N = 6$ with an initial pulse width of $T_{\text{in.FWHM}} = 1$ ps at a wavelength of 1.5 μm is split at the coupler according to:

$$U_c(0, t) = \sqrt{\alpha}\, U(0, t) \tag{7.16}$$

$$U_{cc}(0, t) = i\sqrt{1 - \alpha}\, U(0, t). \tag{7.17}$$

If the incident pulse duration is much shorter than the time taken to transverse the loop length and $1/R$, where R is the pulse repetition rate, the interaction between counter-propagating fields can be neglected. In our entire simulation of a DD-PCF NOLM for the compression of higher-order solitons, we have considered a pulse repetition rate of 50 MHz. Hence, we neglect the interaction between the counter-propagating pulses in the simulation. Similarly, the influence of higher-order effects should be included when analyzing input pulse widths of less than 10 ps. With the inclusion of higher-order effects, the evolutions of U_c and U_{cc} have been calculated by solving the nonlinear Schrödinger equation (NLSE) using the split-step Fourier method (SSFM). Since the loop is asymmetric, the compressions applied to the counter-propagating pulses are different, and the two pulses acquire different phase shifts when they propagate through the fiber. Figures 7.3(a) and (b) show the temporal evolution of the counter-propagating pulses along the fiber length. Here, the optimum loop length is found to be $L = 29$ cm and the compression factor is 30. At the optimum loop length, the switching condition is satisfied for the central peak but not for the rest of the pulse, leading to a low-pedestal compressed pulse. The optimum loop length is chosen such that the compressed pulse approaches a hyperbolic secant pulse, i.e. the pedestal of the compressed pulse is the smallest. In the higher-order soliton compression, the instantaneous power of the central part of the compressed pulse is much higher than that of the pedestals. Therefore, there an enhancement in nonlinearity occurs as the pulse propagates through the fiber, which leads to an enhancement in SPM and effective pulse compression. From the above figures we can clearly see the more effective compression of the clockwise-propagating pulse than that of the counterclockwise-propagating pulse due to the decreasing dispersion experienced by the successively compressed clockwise-propagating pulse. Figures 7.3(c) and (d) represent the spectral profiles of the counter-propagating pulses as they propagate through the fiber. These figures illustrate the effect of wavelength broadening as the pulses propagate through the fiber. The wavelength broadening is greater in the clockwise-propagating pulse than in the

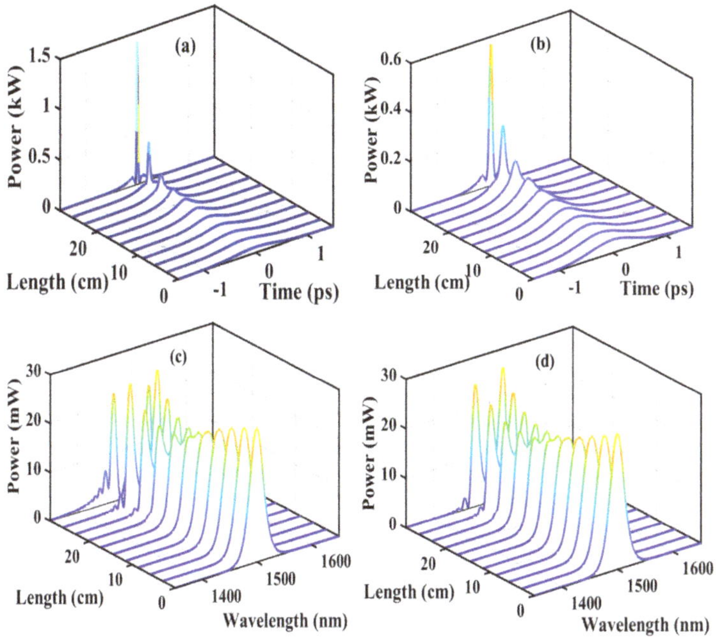

Figure 7.3. Pictorial representation of the temporal (a, b) and spectral (c, d) evolutions of the clockwise-propagating and counterclockwise-propagating pulses.

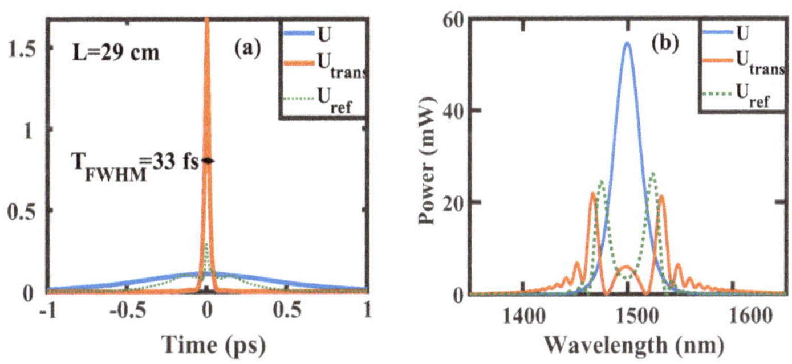

Figure 7.4. Pictorial representation of the temporal (a) and spectral (b) profiles of the final transmitted and reflected pulses.

counterclockwise-propagating pulse because of the enhanced SPM effect experienced by the pulse as the peak power increases along the fiber length. Following recombination of the counter-propagating pulses at the coupler, the final transmitted and reflected pulse profiles in the temporal and spectral domains after recombination at the coupler are presented in figures 7.4(a) and (b). This figure illustrates that due to the nonlinear transmission characteristic of the unbalanced NOLM, the transmissivity of the central part of the compressed pulse is high and the transmissivity of the pedestal is low. As a result, the pedestal of the compressed pulse is suppressed and a low-pedestal narrow pulse is obtained with a FWHM of 33 fs and a peak

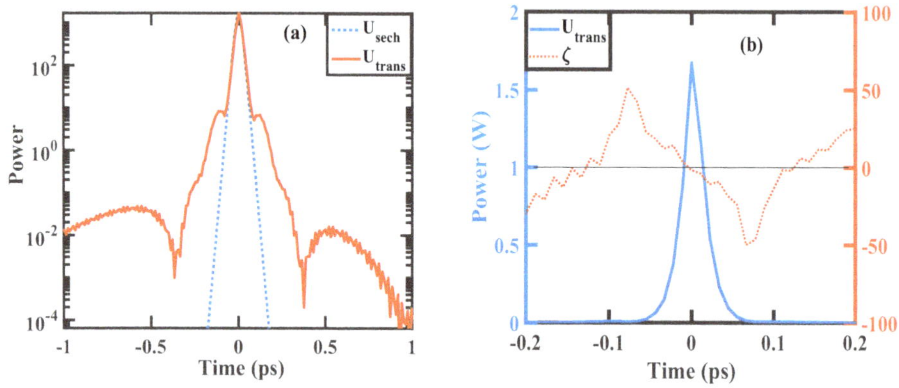

Figure 7.5. (a) Comparison of the log profiles of a transmitted pulse and a hyperbolic secant pulse that have the same pulse width and peak power as that of the compressed pulse, (b) final chirp profile of the transmitted pulse.

power of 1.67 kW. From the spectral profile of the pulse, we can observe that as the peak power increases, both counter-propagating pulses in the NOLM split due to SPM. The notch around the central region is increased due to the frequency chirps around the compressed pulse. The splitting of the pulses influences pedestal suppression. As a result, although the energy transmissivity is relatively high, the pedestal suppression of the pulse from the NOLM is still nonideal. In an efficient compression scheme, the energy contained in the compressed soliton should be equal to that of the input soliton. Hence, we have calculated the energy of the compressed pulse obtained from this compression scheme, which is equal to 5.9 nJ. The calculation of pedestal energy predicts that 6% of the total compressed pulse energy is wasted at the pulse pedestal, which is approximately equal to 0.4 nJ. By appropriately optimizing the loop parameters such as coupler ratio, loop length, and initial chirp parameters, the efficiency of the compression scheme can be further enhanced, as discussed in the following sections. The compressed pulse profile is compared with a hyperbolic secant profile that has the same pulse width and peak power on a logarithmic scale in figure 7.5(a). The compressed pulse profile is very close to the hyperbolic secant profile. From this figure, we can clearly see the presence of a pedestal at the wings of the compressed pulse. Figure 7.5(b) shows the chirp profile of the transmitted pulse after recombination at the coupler. In the central pulse region the chirp value reaches a minimum at around 5 ps^{-2} and varies linearly. The oscillatory behavior in the pulse profile arises due to the pulse pedestal at its wings.

7.6.2 Effects of initial soliton order

The preceding results showed the compression of a 6th-order soliton at a loop length of 30 cm. Since the quality of the compressed pulse is greatly influenced by the input soliton order, the effect of a varying incident soliton order is investigated in this subsection, along with a quality analysis of the compressed pulse using, for example,

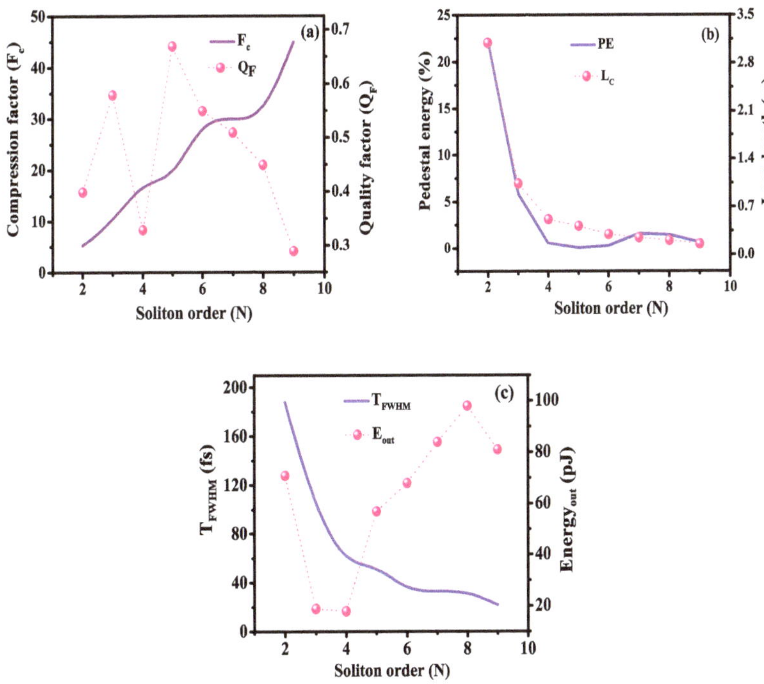

Figure 7.6. Quality analysis of the compression scheme for varying initial soliton order. Variation in (a) compression factor and quality factor, (b) pedestal energy and optimum loop length, (c) FWHM duration of the compressed pulse and output pulse energy.

the compression factor, pedestal energy, quality factor, and optimum loop length. The DD-PCF in each case has the same decreasing rate of dispersion as in the previous analysis, except that the length of the DD-PCF varies due to the impact of the initial soliton order and the coupler's split ratio. For all values of N studied here, the incident solitons have the same initial width of $T_{\text{FWHM}} = 1$ ps and an effective compression ratio of 2.6. The variation in pulse quality parameters for the incidence of different initial soliton orders is represented in figure 7.6. Figure 7.6(a) shows the increase in compression factor with increasing soliton order N, while the quality factor decreases. A maximum quality factor of 0.67 is achieved for an initial soliton order of five. Figure 7.6(b) depicts the change in pedestal energy and optimum loop length. The pedestal energy is very small, i.e. in the range of less than 2%, for $N > 4$. A very small pedestal value is observed for an initial soliton order of five. Thus, the compression is more suited to incident solitons with larger soliton orders. In the higher soliton orders, the pulse quality is influenced by soliton fission, but the higher orders help to enhance the compression factor. In addition, the optimum fiber loop length decreases with increasing soliton order. Figure 7.6(c) portrays the variation in the FWHM of the compressed pulse and its output energy. Here, up to the eighth soliton order, the pulse energy increases; beyond this point, it starts to decrease, and due to the enhanced soliton fission, most of the pulse energy is reflected back into the DD-NOLM compression system.

7.6.3 Effect of initial frequency chirp

In general, the light pulses produced by some lasers, such as directly modulated semiconductor lasers, are inherently chirped. The propagation of such optical pulses through a nonlinear optical medium either enhances or decreases the compression efficiency of the pulse compressor. Hence, this subsection explored the effect of initial frequency chirp on DD-NOLM soliton compression. A linearly chirped incident pulse is represented by:

$$U(0, t) = \sqrt{P_0}\ \operatorname{sech}(t/T_0)\exp(-iCt^2/2), \tag{7.18}$$

where C is the initial chirp parameter. A comparison of the compressed pulse shapes for various chirp parameters such as $C = -5$, $C = 0$, and $C = 5$ is shown in figure 7.7. In each case, the incident pulse has the same peak power, initial pulse width, and an input soliton order of six with a fixed effective compression ratio of 2.6, and the loop length is varied in such a way as to obtain the minimal pedestal for the compressed pulse. From figure 7.7 we can see the variation in compressed pulse temporal duration and peak power for various values of the chirp parameter. The results show that more efficient compression is achieved for negative chirp than for positive chirp. From figure 7.7(a) we can clearly see that most of the pulse energy is reflected back into the fiber. This is because of the enhancement of the SPM effect due to the

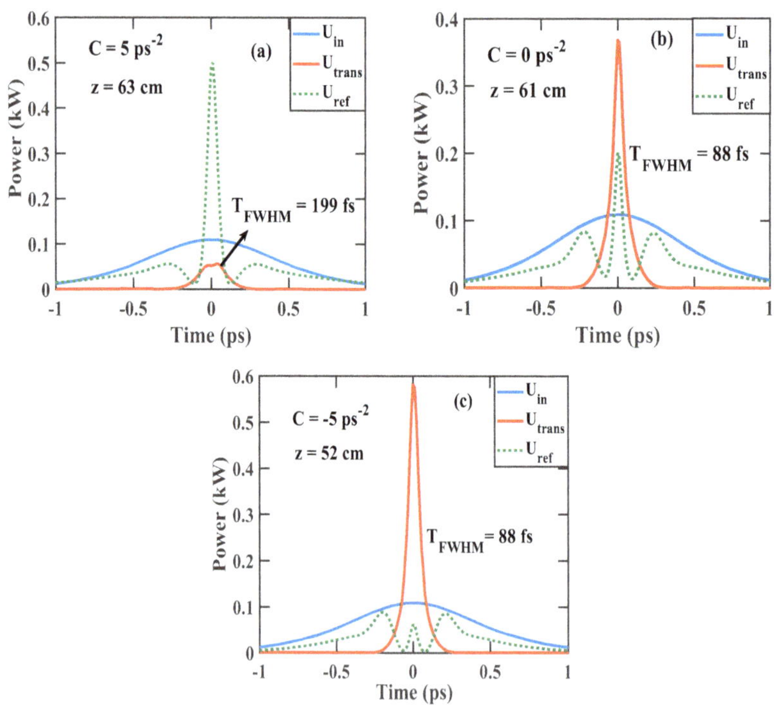

Figure 7.7. Final transmitted pulse profiles for various chirp values (a) $C = 5\ \mathrm{ps}^{-2}$, (b) $C = 0\ \mathrm{ps}^{-2}$, and (c) $C = -5\ \mathrm{ps}^{-2}$.

enhancement in the peak power of both counter-propagating pulses. Hence, upon recombination, due to their different phases, most of the pulses cancel out and are reflected back into the fiber. When compared with figure 7.7(a), in figures 7.7(b) and (c) one can observe an enhancement in compression quality as the chirp value decreases. The optimum fiber loop length also decreases with a decrease in the chirp parameter value. Comparing figures 7.7(b) and (c) clearly shows that even though the compression factors are almost equal for both the zero chirp case and the negative chirp case, for the negative chirp, the values of peak power and output pulse energy are high and a very low pedestal energy can be obtained using an optimally short loop length.

Although the pulses are compressed efficiently while propagating through the loop, in the proposed soliton compression scheme, the percentage of transmittance depends on the sign of the initial chirp parameter. Most of the published reports show that linear frequency chirp can be compensated by a linear dispersive element such as a grating pair, a prism pair, or a chirped fiber Bragg grating. However, the inclusion of these components increases system complexity, causes energy loss, and may alter the pulse shape. Third-order dispersion and Raman self-scattering, for example, have a significant impact on the performance of a fiber Bragg grating. Instead, the utilization of a DD-NOLM as a pulse compressor yields good-quality compressed pulses, even without precompensation of the initial frequency chirp.

Quality analysis of the compression scheme for various chirp parameters shows that a maximum compression factor of 11.2 can be obtained with minimum pedestal energies of 14.7% for the zero chirp case and 5.2% for the case of negative chirp. A minimum pedestal of 2.1% has been observed for the case of positive chirp, however, it only provides a compression factor of five. Similarly, the quality factor is also low for the case of positive chirp, due to the low percentage of transmittance. The results of varying the compression factor, quality factor, pedestal energy, and optimum loop length are shown in figure 7.8 for different input chirp parameters. The presence of negative chirp yields efficient compression with an output FWHM of

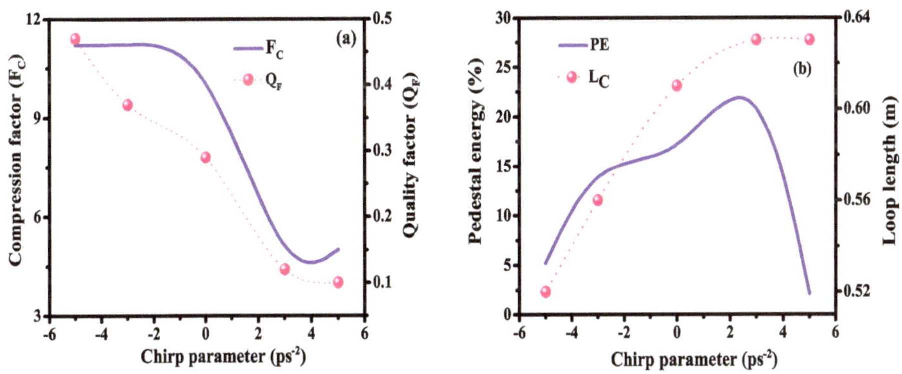

Figure 7.8. Variation in compressed pulse (a) compression factor, quality factor, (b) pedestal energy, and fiber loop length for different initial chirp parameters.

88 fs, an energy of 61 pJ, and a peak power of 582 W for an incident 1 ps FWHM pulse with an energy of 123 pJ and a peak power of 108 W.

7.6.4 Influence of higher-order effects

When investigating the propagation of subpicosecond soliton pulses, the contributions of higher-order phenomena such as RSS and TOD become critical, because soliton compression necessitates a fiber whose length corresponds to multiple soliton periods. The existence of higher-order effects causes oscillations in the neighbourhood of the compressed pulse, which may result in pedestal development and a drop in the compression factor. Hence, this subsection investigated how higher-order effects affect soliton compression in a DD-NOLM. Analysis was performed for optimal pulse compression for cases (i) with RSS and TOD, (ii) without RSS and TOD. In each case, the incident soliton order was the same (N = 6) and T_{FWHM} = 1 ps, and the effective compression ratio was fixed at 2.6. The self-steepening effect was neglected because it plays a much smaller role as compared with RSS and TOD. The changes in pulse shape for the three abovementioned cases are shown in figures 7.9(a) and (b). The results are rather unexpected, because the compression appears to be relatively resistant to higher-order effects, even when the compressed pulse is narrower than 100 fs. In contrast to adiabatic and soliton compression techniques, in which TOD creates oscillations around the compressed pulse, no oscillations arise in this instance, because they are reflected by the DD-NOLM. The most intriguing characteristic is that the presence of RSS and TOD in this case helps to boost DD-NOLM performance. The following describes how RSS improves soliton compression. Due to the asymmetric loop dispersion, counter-propagating pulses encounter varying RSS in the loop, resulting in a relative arrival time delay at the coupler. The overlap region between the two pulses becomes narrower as the time delay grows, compared to the case in which RSS is not included. As a result, the region for which the switching condition is met decreases, resulting in a shorter transmitted pulse. The RSS has little influence on optimal loop length, but has a significant effect on the compression factor and the pedestal energy. When the incident soliton's pulse width is too short, compression enhancement suffers because the temporal separation of the counter-propagating pulses is so great that the two pulses virtually dissociate before adequate compression is applied and

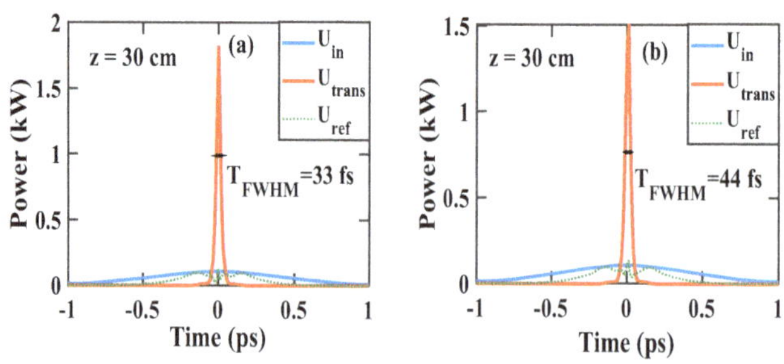

Figure 7.9. Final transmitted pulse profiles (a) with RSS and TOD, (b) without RSS and TOD.

the switching condition is achieved. As a result, for a given input soliton order and a specified loop structure, there is an ideal beginning soliton width for which the RSS-generated compression improvement is maximized.

References

[1] Doran N J and Wood D 1988 Nonlinear optical loop mirror *Opt. Lett.* **14** 56–8

[2] Blow K J, Doran N J and Nayar B K 1989 Experimental demonstration of optical soliton switching in an all-fiber nonlinear Sagnac interferometer *Opt. Lett.* **14** 754–6

[3] Islam M N, Sunderman E R, Stolen R H, Pleibel W and Simpson J R 1989 Soliton switching in a fiber nonlinear loop mirror *Opt. Lett.* **14** 811–3

[4] Wai P K A and Cao W-H 2004 Self-switching of optical pulses in gain-distributed nonlinear amplifying fibre loop mirror *Electron. Lett.* **40** 1208–10

[5] Feng X, Tam H-Y, Liu H and Wai P K A 2006 Multiwavelength erbium-doped fiber laser employing a nonlinear optical loop mirror *Opt. Commun.* **268** 278–81

[6] Fermann M E, Haberl F, Hofer M and Hochreiter H 1990 Nonlinear amplifying loop mirror *Opt. Lett.* **15** 752–4

[7] Steele A L 1993 Pulse compression by an optical fibre loop mirror constructed from two different fibers *Electron. Lett.* **29** 1972–4

[8] Zhao L M, Bartnik A C, Tai Q Q and Wise F W 2013 Generation of 8 nJ pulses from a dissipative-soliton fiber laser with a nonlinear optical loop mirror *Opt. Lett.* **38** 1942–4

[9] Agrawal G 2002 *Fiber Optic Communication Systems* 3rd edn (New York: Wiley)

[10] Smith N J and Doran N J 1995 Picosecond soliton transmission using concatenated nonlinear optical loop-mirror intensity filters *J. Opt. Soc. Am.* B **12** 1117–25

[11] Khrushchev I Y, White I H and Penty R V 1998 High-quality laser diode pulse compression in dispersion-imbalanced loop mirror *Electron. Lett.* **34** 1009–10

[12] Smith K, Doran N J and Wigley P G J 1990 Pulse shaping, compression, and pedestal suppression employing a nonlinear-optical loop mirror *Opt. Lett.* **15** 1294–6

[13] Steele A L 1993 Pulse compression by an optical fibre loop mirror constructed from two different fibres *Electron. Lett.* **29** 1972–4

[14] Chusseau L and Delevaque É 1994 250 -fs optical pulse generation by simultaneous soliton compression and shaping in a nonlinear optical loop mirror including a weak attenuation *Opt. Lett.* **19** 734–6

[15] Wu J, Li Y, Lou C and Gao Y 2000 Optimization of pulse compression with an unbalanced nonlinear optical loop mirror *Opt. Commun.* **180** 43–7

[16] Lidiya A E, Raja R V J and Husakou A 2020 Pulse compression and pedestal suppression by self-similar propagation in nonlinear optical loop mirror *Opt. Commun.* **474** 126083

Chapter 8

Cascaded compression

This chapter discusses the details of a unique cascaded soliton compression approach that uses a chloroform-filled cascaded PCF for the generation of low-energy, few-cycle laser pulses. Higher-order soliton compression is used for the compression of femtosecond pulses.

8.1 Cascaded compression

Following the development of PCFs, numerous experiments based on the simulation results were carried out to demonstrate various pulse-compression techniques using soliton effects. For instance, Tognetti *et al* reported sub-two-cycle pulse generation with a compression factor of nearly 27.02 at $\lambda = 800$ nm using a soliton self-compression technique in a 5 mm PCF that had a completely flat dispersion profile [1]. Amorim *et al* experimentally demonstrated the direct generation of 4.6 fs sub-two-cycle low-energy optical pulse from a sub-nanojoule, 41 fs pulse obtained by soliton-effect compression using a highly nonlinear PCF without additional phase compensation schemes [2]. Recently, Heidt *et al* demonstrated high-quality pulse compression at $\lambda = 800$ nm using an all-normal dispersion PCF set up with a compression factor value of three [3]. Apart from silica-based fibers, soliton pulse compression has also been studied in fibers with various materials and designs. For example, a transform-limited pulse at $\lambda = 1.55\ \mu$m with a duration of 6 ps was compressed to 420 fs in a chalcogenide fiber using numerical simulation [4]. Similarly, using highly nonlinear fiber, Voronin *et al* demonstrated a soliton compression scheme that generated a few single-cycle pulses that had a 2 fs pulse width with a compression ratio of 50 at a central wavelength of 1070 nm using numerical simulation [5]. In this research field, single-cycle and even shorter pulses have been reported for gas-filled and kagome-type hollow-core photonic bandgap fibers using soliton compression schemes [6, 7]. Although several compression techniques can provide compression for pulses as short as sub-cycle pulses, the need for pulse-compression techniques that provide a very high compression factor with low pedestal energy still persists. An analysis of the importance of the quality of the compressed pulses has been performed by González-Baquedano *et al* [8, 9].

Very recently, another mechanism has been developed to generate ultrashort pulses by cascading two different fibers [10–13], which has created profound interest among researchers. Cascading PCFs for pulse compression offers a very high compression factor due to the compression of an already compressed input pulse. In other words, the original input pulse that is compressed at the end of first stage is compressed again in the subsequent stages, resulting in multiple compression of the input pulse. In 2015, Mak *et al* [10] proposed a two-stage kagome PCF compression system. In this method, the first stage in the compression of 250 fs pulses to 22 fs results from the use of a fiber mirror for compression; secondly, compression to 9.1 fs is achieved using soliton-effect compression. With properly managed dispersion and using the nonlinearity of cascaded gas-filled hollow-core fibers, Voronin *et al* theoretically analyzed the generation of sub-100 GW pulses that had a pulse width of about 2 fs from millijoule picosecond laser pulses [11]. Few-cycle cascaded compression had been numerically reported with quadratic phase matching (QPM) using multiple sections of lithium niobate (LN) crystal and two-section bonded β—barium borate (BBO) crystal to achieve a compression factor of 13.88 and a quality factor of 0.72 [12]. Unlike the use of such nonlinear crystals, cascading PCFs for pulse compression has two major factors that have been identified as affecting the quality of the compressed pulse to a much greater extent. The first one is splice loss, which reduces the compression factor significantly when two different fibers are coupled, and the second is the intermediate optical components in the compression system, which induce additional contributions to the pedestal energy of the compressed pulse. However, the noteworthy fact is that cascading PCFs with different dispersion and nonlinear parameters without splicing or using any additional components could provide a very high compression factor and reduced pedestal energy. Hence, a cascaded compression scheme has been proposed that employs a single PCF. In order to benefit from the cascading phenomenon, the fiber parameters are varied using different constant temperatures along the length of the fiber. In this way, the fiber-splice loss does not occur. This chapter further discusses an extensive study of the use of cascaded PCFs to generate ultrashort pulses at shorter wavelengths near 800 nm using the effect of temperature.

8.2 Effect of temperature on chloroform-infiltrated PCF

In order to study the dynamics of propagating solitons in the near-visible regime, the effect of temperature on the PCF parameters was first studied by assuming that the wavelength of 800 nm lies within the anomalous dispersion regime. Unlike typical PCFs, liquid-infiltrated PCFs have a central air hole, which is filled with a specific liquid based on the requirement of the particular application; this relies on the fact that the infiltrating liquid used in any of the PCF air holes has well-defined experimental feasibility [14]. Generally, CS_2 liquid is used to obtain high nonlinearity [15], but it is not suitable for tuning anomalous dispersion in the visible and near-infrared regions, unlike chloroform liquid. In addition to this, chloroform liquid also has high nonlinearity as well as a high thermo-optic coefficient. For these reasons, the hollow core of the fiber is filled with chloroform liquid and the resulting PCF is known as chloroform-infiltrated PCF (CPCF), as shown in figure 8.1. After several numerical simulations and dispersion calculations, the suitably customized and finalized CPCF structure has an air-hole diameter of 0.9 μm with a pitch value of 1.25 μm, creating a high dispersion value at 800 nm. To study the temperature-dependent propagation characteristics of the CPCF, the corresponding temperature- and wavelength-dependent refractive indices of silica and chloroform were employed in the numerical simulation. The refractive index of chloroform liquid at 20 °C is calculated using Sellmeier's equation below, as described in [16]:

$$n^2_{\text{chloro, 20}}(\lambda) = 1.431\,364 + 5632.41/\lambda^2 - 2.0805 \times 10^8/\lambda^4 + 1.2613 \times 10^{13}/\lambda^6, \tag{8.1}$$

where $n_{\text{chloro,20}}(\lambda)$ is the refractive index of chloroform liquid at a temperature of 20 °C as a function of the wavelength, λ. Using the refractive index given by equation (8.1) at T_0 (T = 20 °C), the refractive indices of chloroform at other temperatures T are calculated using the formula [17],

$$n_{\text{chloro}}(\lambda, T) = n_{\text{chloro, 20}}(\lambda) + (T - T_0)\frac{dn}{dT}. \tag{8.2}$$

The thermo-optic coefficient (dn/dT) of chloroform liquid [17] is -6.328×10^{-4} K^{-1}. The temperature-dependent refractive index of silica is given by Sellmeier's equation in [18].

As the boiling temperature of chloroform is 61 °C [19], in order to achieve variation in the fiber parameters the temperature was varied from a room temperature of 20 °C as a minimum to 60 °C as a maximum. The effective refractive index (n_{eff}) of the CPCF as a function of wavelength and temperature was then numerically calculated [15, 20] using the fully vectorial effective index method (FVEIM).

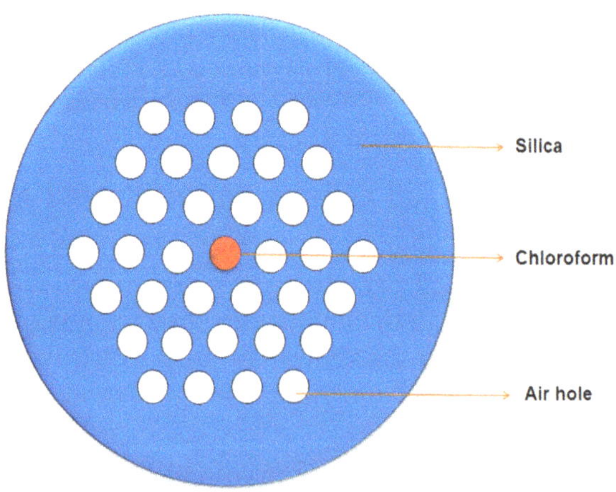

Figure 8.1. Schematic showing the cross-sectional view of the CPCF. Reprinted from [27], with the permission of AIP Publishing.

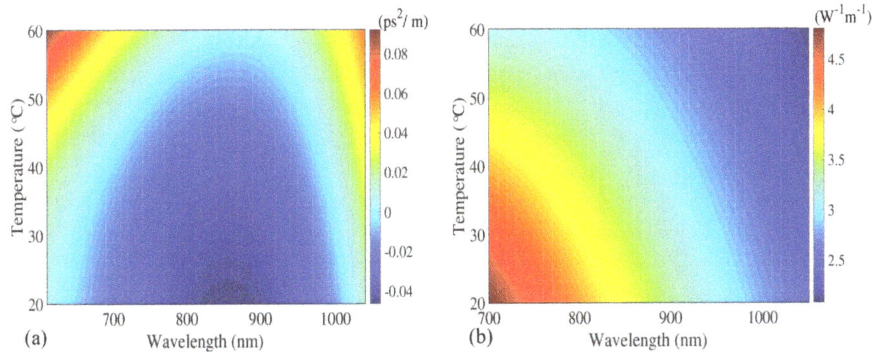

Figure 8.2. Variation of (a) second-order dispersion β_2 and (b) the nonlinear parameter γ as a function of wavelength for different temperatures ranging from 20 °C to 60 °C. Reprinted from [27], with the permission of AIP Publishing.

Using the calculated n_{eff}, the dispersion term is obtained by the approximation of a series expansion, namely $\beta(\omega) \approx \beta(\omega_0) + \sum_n \beta_n(\omega - \omega_0)^n/n!$ around the central frequency ω_0. The change in n_{eff} in turn affects the group velocity dispersion (GVD), the second-order dispersion coefficient β_2, the third-order dispersion coefficient β_3, and the fourth-order dispersion coefficient β_4 of the proposed CPCF. The nonlinear parameter is then calculated using $\gamma(\omega) = \omega n_2/cA_{\mathrm{eff}}$, where n_2 is the nonlinear refractive index (whose value is $1.7 \times 10^{-19}\,\mathrm{m}^2\,\mathrm{W}^{-1}$ for chloroform liquid [16]) and A_{eff} is the effective area of the CPCF. Figures 8.2(a) and (b) show the density plots for β_2 and γ as a function of temperature and wavelength, where we can see that any small variation in the operating temperature of the CPCF affects the dispersion and the nonlinear parameters of the entire CPCF, and hence the dynamics of pulse propagation and pulse compression in the proposed CPCF.

8.3 Theoretical modeling of cascaded PCF

With the aim of investigating its pulse-compression performance, we initially designed the cascaded CPCF to generate ultrashort pulses. It should be noted that a much greater compression factor can be achieved when cascading two or more optical fibers with different dispersion and nonlinear profiles. From the last section, one can observe that the dispersion and nonlinear characteristics of the CPCF can be altered by changing the temperature. Hence, instead of using different PCFs for cascading, this cascaded compression scheme just considers one CPCF and different segments of the fiber which are exposed to different temperatures. This study considers three segments along the cascaded CPCF. Thus, the temperature of the mth segment is denoted by T_m, with m = 1, 2, 3. Figure 8.3 shows a three-dimensional schematic of a chloroform-infiltrated temperature-controlled PCF for better understanding. Hence, the variation in the values of fiber parameters is only achieved by changing the temperature along the fiber span and there is no need to splice different fibers to support multiple pulse compression. This theoretically proposed idea of using temperature-based control along the length of the fiber is practically possible. For instance, Fokine [21] constructed a miniature oven that had

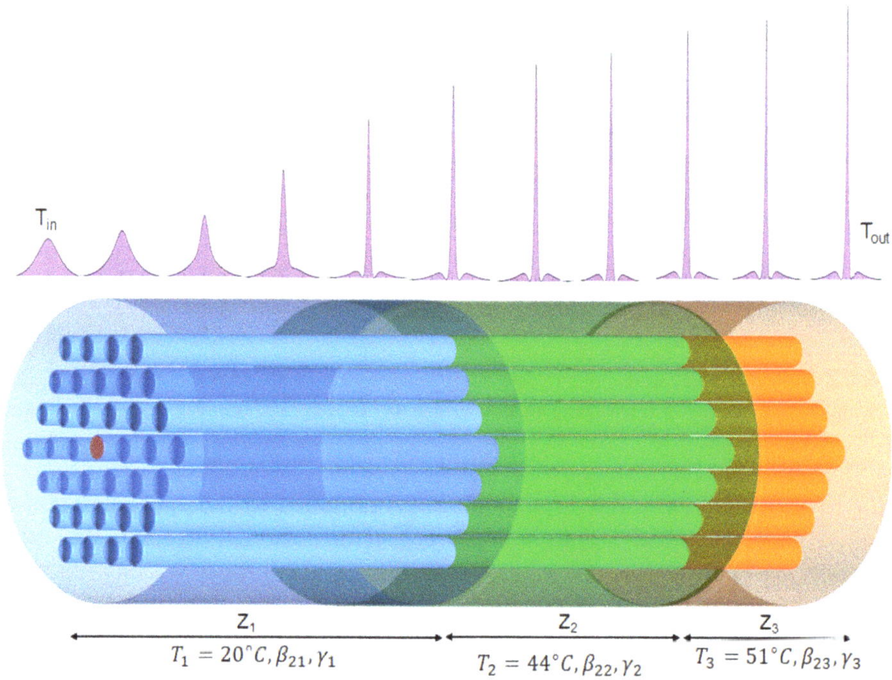

Figure 8.3. Schematic three-dimensional representation of the designed cascaded CPCF compressor along with the qualitative evolution of the pulse propagation. Three different temperatures T_1, T_2, and T_3 are maintained in a single fiber at the lengths of Z_1, Z_2, and Z_3, respectively. Hence, each segment has different fiber parameter values β_{2m} and γ_m. Reprinted from [27], with the permission of AIP Publishing.

distinct heating elements for processing chemical-composition fiber Bragg gratings. It had variable resistors coupled to separate heating elements inside the oven that could account for very low temperature gradients or step-like variations of temperature over distances of less than a millimeter along the fiber.

The central wavelength of 800 nm was used as the input wavelength, for which laser sources such as Ti:sapphire are available to facilitate soliton-driven pulse compression. To achieve a high compression factor within a short distance, the maximum possible dispersion value in the first segment was obtained by choosing the room temperature of 20 °C. In the second and third segments, the temperatures selected were such that the dispersion values were reduced to approximately half the value of the previous segment, as can be seen in table 8.1. Hence, 44 °C and 51 °C were the temperatures selected for the operation of the second and third segments, respectively. Thus, we propose a compressor design based on a novel cascaded PCF, where three different fiber segments of length Z_m on a single CPCF are exposed to different temperatures $T_1 = 20$ °C, $T_2 = 44$ °C, and $T_3 = 51$ °C. This provides different values for the fiber parameters β_{2m} and γ_m at 800 nm for each fiber segment. Table 8.1 shows the corresponding resultant variation in dispersion and the nonlinear parameter for a given value of T in each fiber segment. In addition, the temperature changes that we adopted for pulse compression did not affect the filling fraction and

Table 8.1. Dispersion and nonlinear parameter values for the *m*th segment of the CPCF, where $m = 1, 2, 3$ represent the first, second, and third segments, respectively.

T_m (°C)	$\beta_{2m} \times 10^{-3}$ (ps^2m^{-1})	$\beta_{3m} \times 10^{-5}$ (ps^3m^{-1})	$\beta_{4m} \times 10^{-7}$ (ps^4m^{-1})	γ_m ($W^{-1}m^{-1}$)
20	−42.0517	3.3924	1.4297	4.0524
44	−21.4417	3.3762	1.5892	3.0214
51	−10.4991	3.9191	1.736 25	2.6910

air-hole pitch of the microstructured CPCF, as the boiling temperature of silica is 1800 °C. Silica-based fibers are usually drawn at the temperature of 1950 °C (well above its melting temperature) for fine structuring [22].

The evolution of soliton propagation through different segments of a cascaded PCF has been studied by solving the modified nonlinear Schrödinger equation (MNLSE) [23] with the inclusion of higher-order dispersion, Kerr nonlinearity, and intrapulse stimulated Raman scattering, as follows:

$$\begin{aligned} &\frac{\partial U}{\partial z} - \sum_{n=2}^{4} \beta_{nm} \frac{i^{n+1}}{n!} \frac{\partial^n U}{\partial t^n} \\ &= i\gamma_m \left[1 + \frac{i}{\omega_0} \frac{\partial}{\partial t}\right] \left[U(z, t) \int_0^{\infty} R(t')|U(z, t - t')|^2 \, dt'\right], \end{aligned} \tag{8.3}$$

where U is the envelope of the propagating pulse, z and t are the propagation distance and propagation time, respectively, and the parameters β_{nm} and γ_m are the dispersion coefficient of the nth order and the Kerr nonlinear parameter at the mth segment of the cascaded CPCF, respectively. $R(t)$ is the response function expressed as $(1 - f_R)\,\delta(t) + f_R\, h_R(t)$; it includes both the instantaneous electronic and delayed Raman contributions with $f_R = 0.35$ [16] and $h_R(t)$ is the Raman response function given by

$$h_R(t) = \frac{\tau_1^2 + \tau_2^2}{\tau_1 \tau_2^2} \exp(-t/\tau_2) \sin(t/\tau_1)\, \Theta(t), \tag{8.4}$$

where $\tau_1 = 38$ fs and $\tau_2 = 220$ fs for chloroform [16]. $\Theta(t)$ is the Heaviside step function and $\delta(t)$ is the Dirac delta function.

8.4 Compression through a cascaded PCF

Dynamics and compression studies were carried out numerically by applying the split-step Fourier method (SSFM) to the MNLSE. For the SSFM studies, a hyperbolic secant pulse was assumed to be the input, which was given by

$$U(0, t) = N_1 \operatorname{sech}(t/T_{in}), \tag{8.5}$$

where N_1 is the soliton order of the first segment $m = 1$. The soliton order of the mth segment is given by $N_m^2 = (\gamma_m P_{in} T_{in}^2)/(|\beta_{2m}|)$, where P_{in} and T_{in} are the input peak power and input pulse width, respectively. Here, $T_{in} = T_{FWHM,\,in}/(2\log(1+\sqrt{2}))$ where $T_{FWHM,\,in}$ is the FWHM of the input pulse. The length L of the CPCF is considered to be the sum of the lengths of all segments and is given by $L = \sum_{m=1}^{3} Z_m$. In order to compress a pulse using the higher-order soliton compression scheme, an input pulse at the central wavelength of 800 nm and initial values of $T_{FWHM,\,in} = 140$ fs and $P_{in} = 10$ W form the input to the first segment $m = 1$. This input pulse can trigger a soliton pulse of non-integer order $N_1 = 2.46$ in the first segment of the fiber. Figure 8.4 shows the spectral and temporal evolution of the propagating soliton pulse along the proposed cascaded CPCF at the first segment. In general, as the higher-order soliton propagates inside the fiber, the pulse shape becomes narrower over a certain length and then broadens to regain its initial shape. Hence, a higher-order soliton pulse-compression scheme usually takes the output pulse at the point of maximum narrowing over the course of its propagation.

The required optimum length $Z_{m-\mathrm{opt}}$, where the pulse attains maximum compression for all $m = 1, 2, 3$ was calculated using [1]

$$Z_{m-\mathrm{opt}} = \frac{\pi}{2}\left[\frac{0.32}{N_m} + \frac{1.1}{{N_m}^2}\right] L_{\mathrm{Dm}}, \tag{8.6}$$

where the dispersion length of the mth segment L_{Dm} is defined as $T^2/|\beta_{2m}|$. Even though the optimum length is calculated using equation (8.6), the exact length required for compression by each segment m will change slightly due to the inclusion of higher-order effects. Hence, the optimum length has been numerically calculated and fixed at 11.5 cm, which is where the maximum compression takes place. It has been calculated that compressed pulses as short as 31.79 fs (FWHM) can be obtained using a pulse energy of 1.25 pJ for segment $m = 1$.

The compressed output pulse from the first segment, which is maintained at the temperature T_1 (20 °C) is then launched into the second segment $m = 2$ at $T_2 = 44$ °C. The β_{22} and γ_2 values of the second segment are shown in table 8.1. Although a constant temperature value was been considered over each particular segment of the fiber, a temperature gradient was also considered for a length of 20 μm at the junction of these two segments, where the temperature changes from 20 °C to 44 °C. Hence, it is assumed that the dispersion and nonlinearity decrease linearly from -42.0517×10^{-3} (ps^{2}m^{-1}) to -21.4417×10^{-3} (ps^{2}m^{-1}) and 4.0524 (W^{-1}m^{-1}) to 3.0214 (W^{-1}m^{-1}), respectively, at the junction. Thus, the MNLSE has been numerically solved by considering the linear variation of dispersion and nonlinearity at the junctions of two PCF segments. As both dispersion and nonlinearity decrease simultaneously within a small length, the shape of the soliton pulse remains the same. Thereafter, the pulse is allowed to propagate in the second segment as shown in figure 8.5. The corresponding soliton order at $m = 2$ changes to $N_2 = 1.34$ and hence the required fiber length changes to Z_2. Thus, the 31.79 fs output soliton pulse from the first segment is allowed to propagate into the second segment, which has the shorter length of 3.0 cm. The higher-order soliton effects, i.e. pulse evolution and compression, in the second segment are

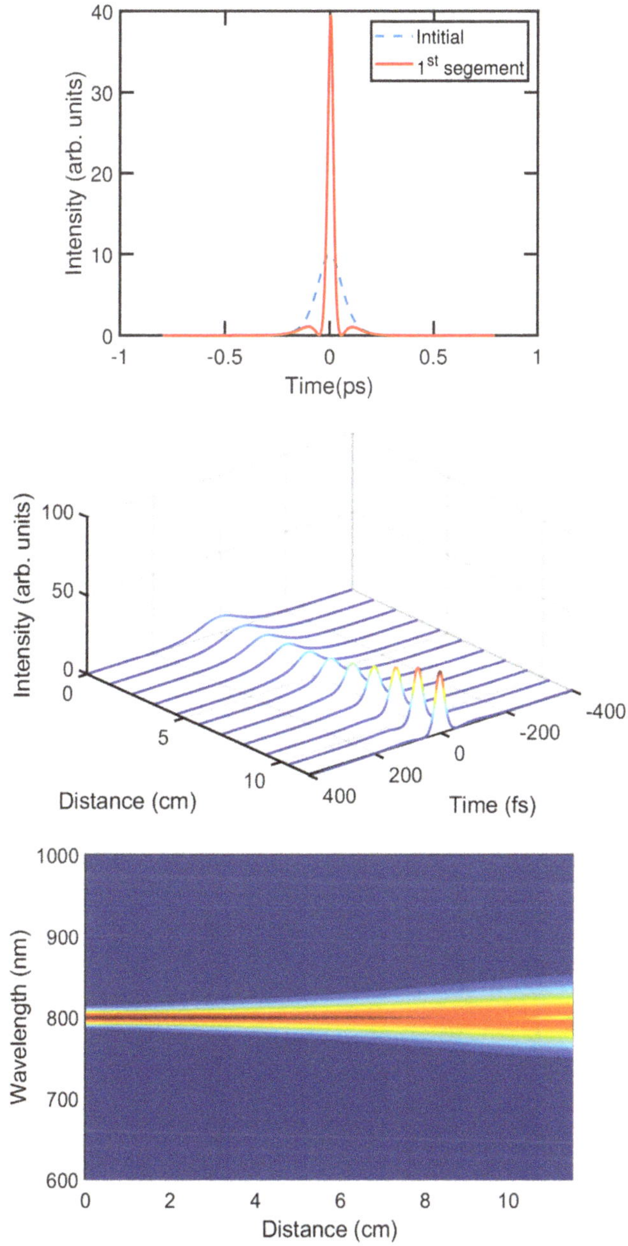

Figure 8.4. The temporal and spectral evolution of higher-order soliton propagation through CPCF over the length of the first segment $Z_1 = 11.5$ cm. The input pulse parameters P_{in} and $T_{\text{FWHM, in}}$ are 10 W and 140 fs, respectively. Reprinted from [27], with the permission of AIP Publishing.

analogous to those in the first segment. The 23.65 fs output pulse from the second segment is then allowed to propagate into third segment, where $T_3 = 51\,^{\circ}\text{C}$ and the temperature variation over the length of 10 μm is considered. The length Z_3 required for soliton order $N_3 = 1.54$ is adjusted from the corresponding β_{23} and γ_3 values at this

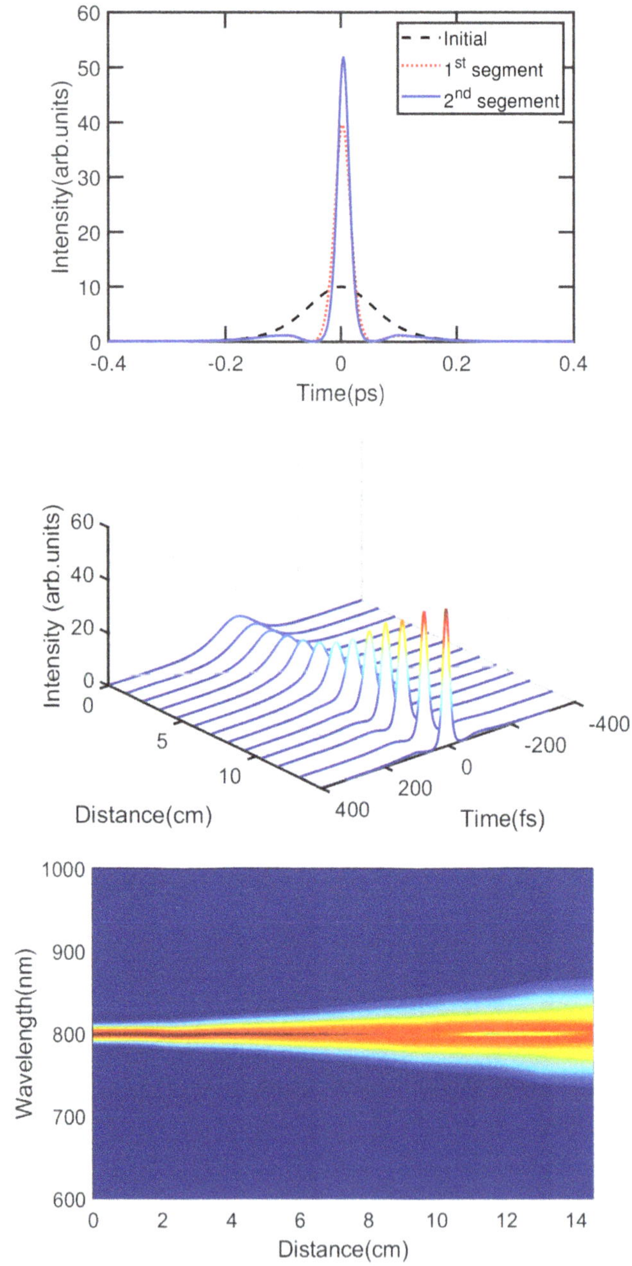

Figure 8.5. The temporal and spectral evolution of higher-order soliton propagation through CPCF over the lengths of the first and second segments $Z_1 + Z_2 = 14.5$ cm. Reprinted from [27], with the permission of AIP Publishing.

temperature and requires a value of 2.9 cm to get the final compressed 12.02 fs output pulse. In figure 8.6, it can be observed that as the soliton pulse propagates along the CPCF and evolves into a higher-order soliton, the generated ultrashort pulses are compressed further as the temperature changes in each segment and this

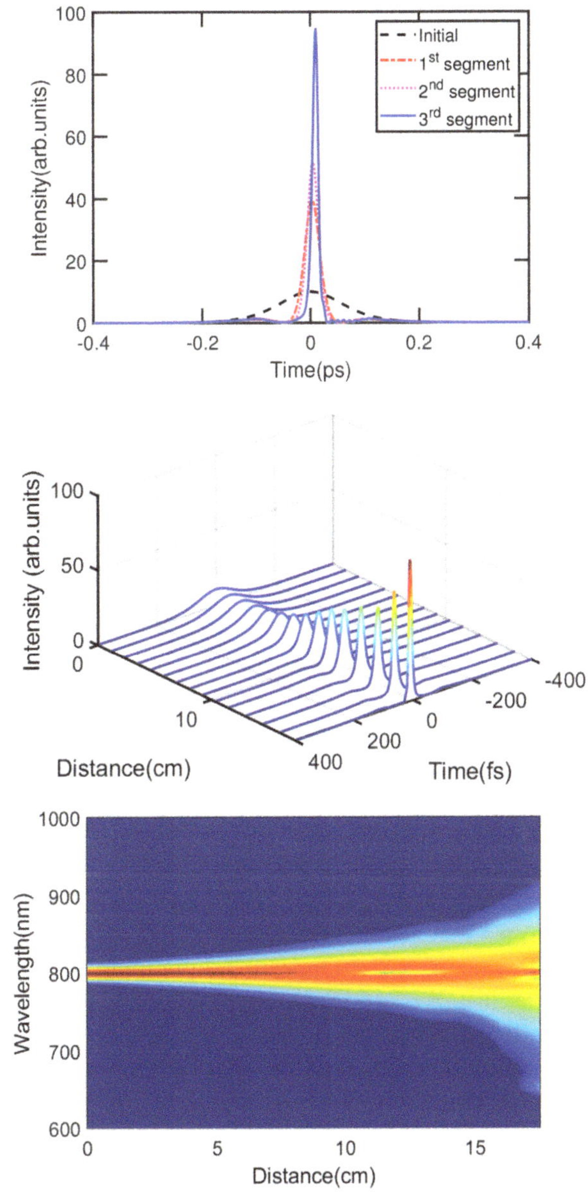

Figure 8.6. The temporal and spectral evolution of higher-order soliton propagation through CPCF over the lengths of the first, second, and third segments $Z_1 + Z_2 + Z_3 = 17.4$ cm. Reprinted from [27], with the permission of AIP Publishing.

pulse-compressing process continues to narrow the pulse width. This chain of soliton compression processes starts with 140 fs input pulses and ends in the generation of short 12.02 fs pulses. Table 8.2 tabulates the values that the pulses take in each segment of the CPCF.

Table 8.2. Output power $P_{\mathrm{out},m}$, pulse width $T_{\mathrm{FWHM,\ out},\ m}$ and soliton-order N_m at the end of each segment of respective length Z_m where $m = 1, 2, 3$. The input pulse has an input power of $P_{\mathrm{in}} = 10$ W and an initial pulse width of $T_{\mathrm{FWHM,\ in}} = 140$ fs.

m	$P_{\mathrm{out},m}$ (W)	$T_{\mathrm{FWHM,out},m}$ (fs)	N_m	Z_m (cm)
1	39.4	31.79	2.46	11.5
2	51.8	23.65	1.34	3.0
3	94.4	12.02	1.54	2.9

8.5 Quality analysis

The influence of input pulse energy on the evolution of higher-order soliton dynamics to generate compressed pulses via cascaded CPCF is discussed in this section. Hence, apart from the $N = 2.46$ soliton compression corresponding to an input pulse power of 10 W, the dynamics of the propagating pulse is analyzed for various soliton orders by tuning the pump power. We analyze the use of a pulse that has a width of 140 fs with pump powers varying from 10 to 70 W in increments of 10 W. In this scenario, the soliton number increases as the input pump power increases and the obtained values are 2.46, 3.48, 4.27, 4.93, 5.51, 6.04, and 6.52, respectively. Furthermore, it is found that, as the dispersion and nonlinear parameters change, the corresponding soliton order N_m varies, hence the optimum segment length Z_m required for compression in each segment m also changes. Figure 8.7 shows the variation of the soliton order in the first segment and the corresponding full-compression length L of the cascaded CPCF for a range of input powers from 10 W to 70 W. It can be observed that the total optimum length of the cascaded CPCF depends on the soliton order through the dependence on P_{in}. For instance, the compression length reduces from 17.4 cm to 2.4 cm when the soliton number is increased from 2.46 to 6.52. When the pump power is greater than 70 W, the soliton order N_1 has values greater than six. Under these pumping conditions, the pulse is compressed at a smaller distance and breaks up by soliton fission due to higher-order effects, Raman scattering, and self-steepening. Therefore, the pump power is limited to 70 W to maintain the pulse shape, and the effect of the pump power on the pulse width of the generated ultrashort pulses is analyzed for the proposed cascaded CPCF.

The major factors that determine the efficiency of the designed compressor are the compression factor and the pedestal energy [24], which are defined in section 5.6. The pulse widths of the ultrashort pulses at the end of third segment for the considered pumping power ranges of 10 W to 70 W are compared by solving equation (8.3). The calculated compression factors and output pulse widths of the compressed pulses are presented in figure 8.8 as a function of pumping power for the soliton compression process in the considered cascaded CPCF.

As a comparative study of the pump power, figure 8.8 illustrates the major impact of the input pulse energy in tuning the pulse width of the generated ultrashort pulses. Considering a pump pulse width $T_{\mathrm{FWHM,\ in}}$ of 140 fs, the pulse width of the

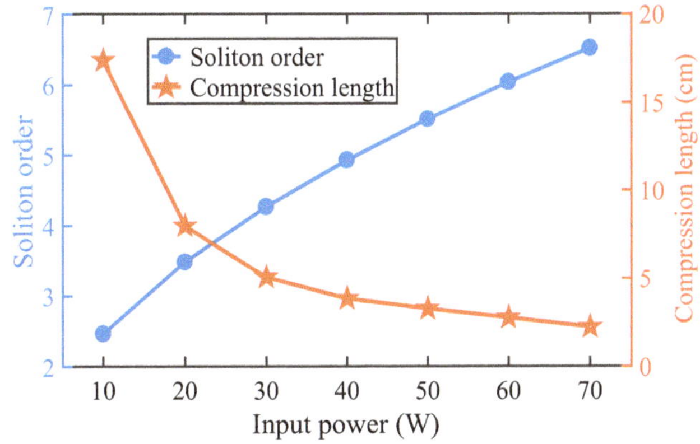

Figure 8.7. Plot of the soliton orders of the first segment and optimum compression lengths as a function of pump power for the designed CPCF compressor. The considered input pulse width is $T_{\text{FWHM, in}} = 140$ fs. Reprinted from [27], with the permission of AIP Publishing.

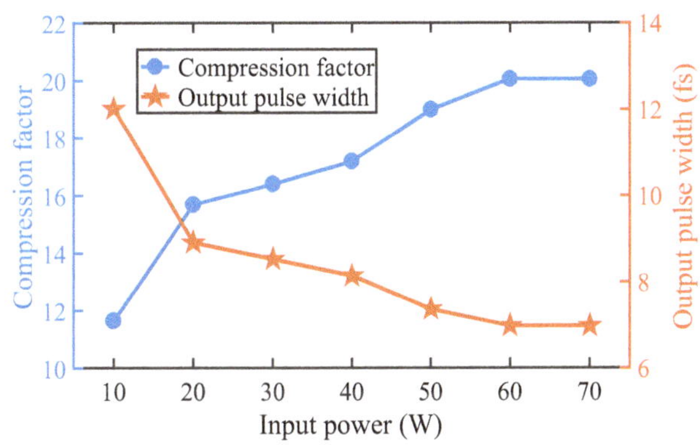

Figure 8.8. Compression factors and output pulse widths as a function of input power from 10 to 70 W. The initial pulse width is $T_{\text{FWHM, in}} = 140$ fs. Reprinted from [27], with the permission of AIP Publishing.

compressed pulse decreases from 12.02 to 6.98 fs for an increase in the pump power from 10 to 70 W, respectively, which means that the compression factor can be increased from 11.64 to 20.05, correspondingly. The corresponding calculated output powers and energies of the compressed soliton pulses for the same $T_{\text{FWHM, in}}$ are shown in figure 8.9. It can be seen that small variations of the pump power directly affect the soliton-order values of the proposed fiber. Such a dependence, in turn, leads to large compression factors and short compressed pulses with high output power and energy in the cascaded CPCF.

To analyze the quality of the compressed pulse, it is also necessary to study the quality factor and pedestal energy of the compressed pulse at the output end of the CPCF. A large quality factor and a minimal pedestal energy characterize high-quality compressed pulses. Due to the inclusion of higher-order dispersion as well as nonlinear effects, it could be difficult to prevent the pulse shape from pulse breaking and shifting, which could generate a large pedestal unless an appropriate pump

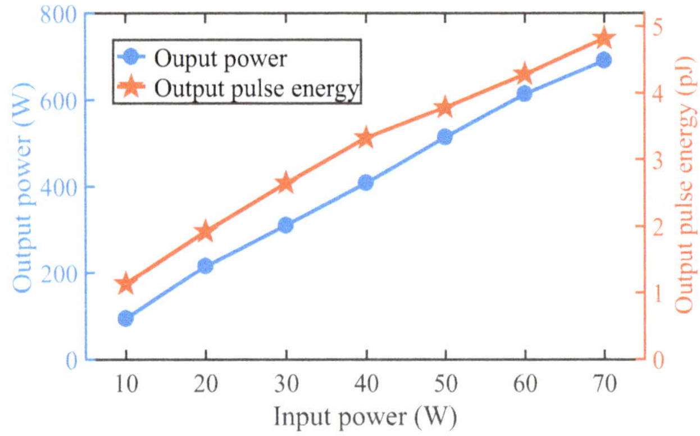

Figure 8.9. Output power and energy as a function of input power from 10 to 70 W. Reprinted from [27], with the permission of AIP Publishing.

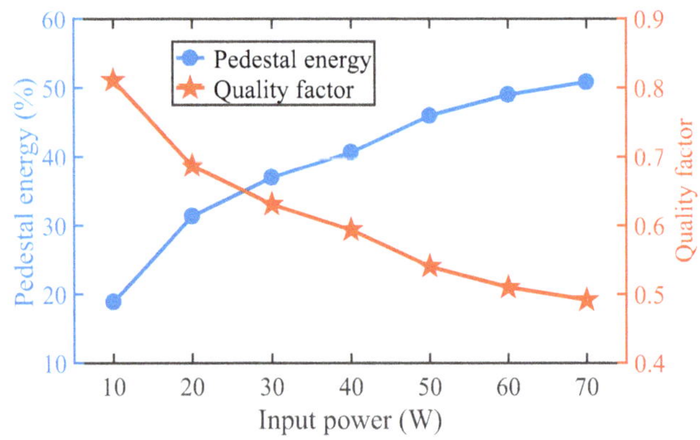

Figure 8.10. Pedestal energy (in %) and quality factor as a function of input power from 10 to 70 W. Reprinted from [27], with the permission of AIP Publishing.

power were to be selected. This can clearly be seen in figure 8.10, which depicts the pedestal energy as well as the quality factor of the compressed pulse as a function of the pumping power. From this figure, it can be seen that the pedestal energy of the compressed pulse increases as the input power increases. For an input pulse with a pump power of 10 W, a quality factor of 0.81 and a pedestal energy of 18.92% were obtained for the output compressed pulse. Meanwhile, for an input pumping power of 70 W, the quality factor decreases to 0.49 and 50.81% of pedestal energy is generated. Thus, as the input pulse power increases, the quality of pulse compression tends to degrade and saturate for input powers greater than 50 W.

The effect of the error due to the gradual variation of temperature between consecutive segments on the pulse compression is estimated in terms of the change in the total photon number [25] at the junction. The change of photon number $\partial P/\partial z$ with distance must be zero in order to maintain energy conservation. Thus, a relative non-zero change in the photon number gives a measure of the numerical error arising from pulse compression. The photon number P is calculated using equation (8.7), as follows:

$$P(z) = \int \frac{|\tilde{U}(z, \omega)|^2}{\omega} d\omega, \tag{8.7}$$

where $\tilde{U}(z, \omega)$ is the Fourier transform of the pulse envelope $U(z, t)$. The percentage changes in the photon number at the first (dph_1) and second (dph_2) junctions with the change in pump power are depicted in figure 8.11. It is numerically calculated that dph_1 is 0.16% and dph_2 is 0.29% for an input power of 10 W, while for an input power of 70 W, the percentage changes in photon number are 0.47% and 0.14% at the first and second junctions, respectively.

Most of the previously proposed cascading higher-order soliton compression schemes worked based on a choice of different fiber couplings together with optical elements to obtain a compressed pulse from compressed input pulses. Instead of using various fibers with different dispersion and nonlinear parameters, we use only one fiber that is exposed to three different temperatures, 20 °C, 44 °C, and 51 °C at three segments along the fiber. Here, the variation of temperature controls both the dispersion and the nonlinear parameters of the entire CPCF. Using the proposed novel cascaded PCF, by varying the temperature, we can generate ultrashort pulses without any external optical elements and splice loss can be avoided. The effect of different temperatures along the length of the fiber helps to achieve the respectable compression of a 140 fs pulse to a 6.98 fs pulse with a low peak output energy of 4.81 pJ when light is launched at the pumping wavelength of 800 nm in 2.4 cm (length) of a CPCF. Using this novel compression scheme, a few-cycle laser pulse with a high quality factor of 0.49 and a pedestal energy of 50.81% has been demonstrated. In table 8.3, we compare the obtained values for the compression factor and quality factor of our CPCF compressor with others previously reported for fiber compressors operating at the same central wavelength of 800 nm. From a 140 fs input pulse and 70 W of pump power, the generation of short 6.98 fs pulses is made possible. The output pulse also has a very low output energy of 4.81 pJ accompanied by 50.81% of pedestal energy. The quality factor is estimated to be 0.49. To our knowledge, the short pulse obtained using this fiber has lower energy

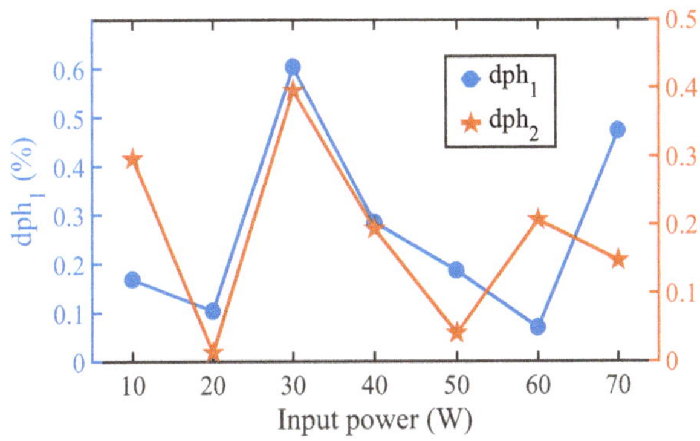

Figure 8.11. The changes in photon number (in %) at first and second junctions as a function of input power from 10 to 70 W.

Table 8.3. Comparison of different fibers operating at a central wavelength of 800 nm for the application of pulse compression. $T_{FWHM,\ in}$ and $T_{FWHM,\ out}$ are the input and output pulse widths, E_{out} is the output pulse energy, Q_c is the quality factor, and F_c is the compression factor.

Ref.	$T_{FWHM,\ in}$ (fs)	$T_{FWHM,\ out}$ (fs)	E_{out}	Q_c	F_c
[26]	70	6.8	0.63 nJ	0.73	10.29
[1]	100	3.7	1.59 nJ	0.32	27.02
[2]	41	4.9	0.1519 nJ	0.24	8.36
[3]	15	5	—	—	3
This work	140	6.98	4.81 pJ	0.49	20.05

and better quality and the CPCF compressor provides a higher compression factor than most of the other compression schemes at a pumping wavelength of 800 nm.

Parts of this chapter have been reprinted from [27].

References

[1] Tognetti M V and Crespo H M 2007 Sub-two-cycle soliton-effect pulse compression at 800 nm in photonic crystal fibers *J. Opt. Soc. Am.* B **24** 1410–5

[2] Amorim A A, Tognetti M V, Oliveira P, Silva J L, Bernardo L M, Kärtner F X and Crespo H M 2009 Sub-two-cycle pulses by soliton self-compression in highly nonlinear photonic crystal fibers *Opt. Lett.* **34** 3851–3

[3] Heidt A M, Rothhardt J, Hartung A, Bartelt H, Rohwer E G, Limpert J and Tünnermann A 2011 High quality sub-two cycle pulses from compression of supercontinuum generated in all-normal dispersion photonic crystal fiber *Opt. Express* **19** 13873–9

[4] Fu L, Fuerbach A, Littler I C M and Eggleton B J 2006 Efficient optical pulse compression using chalcogenide single-mode fibers *Appl. Phys. Lett.* **88** 081116

[5] Voronin A A and Zheltikov A M 2008 Soliton-number analysis of soliton-effect pulse compression to single-cycle pulse widths *Phys. Rev.* A **78** 063834

[6] Balciunas T, Fourcade-Dutin C, Fan G, Witting T, Voronin A A, Zheltikov A M, Gerome F, Paulus G G, Baltuska A and Benabid F 2015 A strong-field driver in the single-cycle regime based on self-compression in a kagome fibre *Nat. Commun.* **6** 1–7

[7] Ouzounov D G, Hensley C J, Gaeta A L, Venkateraman N, Gallagher M T and Koch K W 2005 Soliton pulse compression in photonic band-gap fibers *Opt. Express* **13** 6153–9

[8] González-Baquedano N, Arzate N, Torres-Gómez I, Ferrando A, Ceballos-Herrera D E and Milián C 2012 Femtosecond pulse compression in a hollow-core photonic bandgap fiber by tuning its cross section *Photonics Nanostruct. Fundam. Appl.* **10** 594–601

[9] González-Baquedano N, Torres-Gómez I, Arzate N, Ferrando A and Ceballos-Herrera D E 2013 Pulse quality analysis on soliton pulse compression and soliton self-frequency shift in a hollow-core photonic bandgap fiber *Opt. Express* **21** 9132–43

[10] Mak K F, Seidel M, Pronin O, Frosz M H, Abdolvand A, Pervak V, Apolonski A, Krausz F, Travers J C and Russell P St J 2015 Compressing μJ-level pulses from 250 fs to sub-10 fs at 38-MHz repetition rate using two gas-filled hollow-core photonic crystal fiber stages *Opt. Lett.* **40** 1238–41

[11] Voronin A A, Mikhailova J M, Gorjan M, Major Z and Zheltikov A M 2013 Pulse compression to subcycle field waveforms with split-dispersion cascaded hollow fibers *Opt. Lett.* **38** 4354–7

[12] Zeng X, Guo H, Zhou B and Bache M 2012 Soliton compression to few-cycle pulses with a high quality factor by engineering cascaded quadratic nonlinearities *Opt. Express* **20** 27071–82

[13] Lidiya A E, Raja R V J and Varshney S K 2021 Low pedestal sub-17 fs pulse generation through cascaded self-similar compression in photonic crystal fibers *J. Opt.* **23** 125503

[14] Wang F, Yuan W, Hansen O and Bang O 2011 Selective filling of photonic crystal fibers using focused ion beam milled microchannels *Opt. Express* **19** 17585–90

[15] Zhang R, Teipel J and Giessen H 2006 Theoretical design of a liquid-core photonic crystal fiber for supercontinuum generation *Opt. Express* **14** 6800–12

[16] Raja R V J, Porsezian K, Varshney S K and Sivabalan S 2010 Modeling photonic crystal fiber for efficient soliton pulse propagation at 850 nm *Opt. Commun.* **283** 5000–6

[17] Samoc A 2003 Dispersion of refractive properties of solvents: chloroform, toluene, benzene, and carbon disulfide in ultraviolet, visible, and near-infrared *J. Appl. Phys.* **94** 6167–74

[18] Ghosh G, Endo M and Iwasaki T 1994 Temperature-dependent Sellmeier coefficients and chromatic dispersions for some optical fiber glasses *J. Lightwave Technol.* **12** 1338–42

[19] John A D *et al* 1999 *Lange's Handbook of Chemistry* 15th edn (New York: McGraw-Hill)

[20] Raja R V J and Porsezian K 2007 A fully vectorial effective index method to analyse the propagation properties of microstructured fiber *Photon. Nanostruct. Fundam. Appl.* **5** 171–7

[21] Fokine M 2001 High temperature miniature oven with low thermal gradient for processing fiber Bragg gratings *Rev. Sci. Instrum.* **72** 3458–61

[22] Ballato J *et al* 2008 Silicon optical fiber *Opt. Express* **16** 18675–83

[23] Agrawal G P 2007 *Nonlinear Fiber Optics* 4 (Cambridge, MA: Academic Press)

[24] Hussein R A, Hameed M F O and Obayya S S A 2015 Ultrahigh soliton pulse compression through liquid crystal photonic crystal fiber *IEEE J. Sel. Top. Quantum Electron.* **22** 302–9

[25] Frosz M H, Sørensen T and Bang O 2006 Nanoengineering of photonic crystal fibers for supercontinuum spectral shaping *J. Opt. Soc. Am.* B **23** 1692–9

[26] Foster M A, Gaeta A L, Cao Q and Trebino R 2005 Soliton-effect compression of supercontinuum to few-cycle durations in photonic nanowires *Opt. Express* **13** 6848–55

[27] Yamunadevi R, Vasantha Jayakantha Raja R and Arzate N 2018 Generation of low power and ultrashort laser pulses at 800 nm through soliton compression in chloroform-infiltrated cascaded photonic crystal fibers *J. Appl. Phys.* **124** 113105 113105

Chapter 9

Supercontinuum compression

This chapter theoretically explains the generation of a tunable broadband light spectrum obtained by varying the temperature of a photonic crystal fiber (PCF), which is found to have a significant impact on the process of SCG-induced pulse compression. The temperature-dependent fiber parameters are used to solve the nonlinear Schrödinger equation (NLSE), which determines the evolution of the optical pulse through the proposed fiber model.

9.1 Supercontinuum generation

Most optoelectronics-device-based applications now rely on coherent white-light sources, which span the entire spectral range of the rainbow from violet to blue, green, yellow, orange, as well as red and near-infrared. As a result, several research groups have attempted to develop such white-light sources in recent decades. The phenomenon of generating such a coherent white-light source is known as supercontinuum generation (SCG). SCG describes the process whereby an optical pulse that initially has a narrow spectrum undergoes significant spectral broadening in a nonlinear medium caused by nonlinear optical effects to yield a very bright, coherent, and spectrally continuous output [1, 2]. The SC is spatially coherent and the spectral bandwidth can span several hundreds of nanometers [3, 4]. Alfano and Shapiro were the first to report the generation of a supercontinuum in a bulk medium of borosilicate glass. They observed spectral broadening of a picosecond second-harmonic output of a neodymium garnet laser with an energy of about 5 mJ by laser radiation [5]. In addition, supercontinuums have been generated in conventional fibers; had improved efficiency and spatial properties compared to those generated in bulk media. Lin and Stolen studied SCG in conventional fiber for the first time in 1976 by pumping in the normal group velocity dispersion (GVD) regime [6], and Beaud observed it in the anomalous dispersion regime in 1987 [7]. In these cases, it was discovered that SCG in fiber was primarily caused by soliton propagation, specifically, the breakup of the injected pulse via the soliton fission process.

doi:10.1088/978-0-7503-2686-5ch9

SCG via optical fiber is an interesting method of realizing broadband sources suitable for applications in the vicinity of the 1550nm telecommunications window, particularly in the context of developing wavelength-division multiplexing (WDM) systems, and it motivated extensive research activities in the 1990s [8–10]. The advent of PCF significantly improved supercontinuum technology. SCG using PCF is the technology of choice for the next generation of ultrabroadband coherent light sources. PCFs with a high degree of design flexibility in their micro-structured cladding endowed with significantly tailorable modal properties, such as the ability to engineer dispersion, increased PCF nonlinearity, the ability to move the zero dispersion wavelength (ZDW) to shorter wavelengths, and an adjustable effective mode area, are potential candidates for generating SCs, allowing SCG in PCF to be observed over a much wider range of source parameters than in conventional fibers [1–3]. This has enabled the broadest spectra to be obtained by pumping in the anomalous dispersion regime using a commonplace Ti:sapphire laser centered at around 800 nm. Another benefit of PCF was its ability to tightly confine light to the core, which increased nonlinearity and thus efficiency. Ranka *et al* discovered SCG in PCF by using 100 fs nanojoule energy pulses at 770 nm from a self-mode-locked Ti:sapphire laser to generate a 550 THz bandwidth SC spanning an octave from 400 to 1500 nm through 75 cm of PCF with a ZDW in the region of 765–775 nm [11]. Since then, SCG in PCF has received a great deal of attention for both its fundamental and application aspects, owing primarily to its nonlinear applications in a variety of research fields.

Numerous studies of SCG in PCF have been published in the last decade for all pump regimes ranging from continuous wave (CW), to nanosecond, picosecond, and femtosecond. [12]. Price *et al* observed SCG over 400–1700 nm in a PCF that had a length of 7 m while injecting 350 fs pulses using a Yb^{3+}-doped fiber laser operating at around 1060 nm in the femtosecond regime [13]. A significant amount of work has also been done using Er3+-doped fiber-based SC sources at around 1550 nm. For example, Nishizawa *et al* reported a SC spectrum from 1100 nm to 2100 nm using a femtosecond fiber laser producing 110 fs pulses at 1550 nm [14]. Parallel to these impressive femtosecond results, there has been a lot of interest in generating broadband SC with low-power picosecond and even nanosecond pulses. Niklov *et al* used picosecond pulses to generate a broad SC in a PCF and demonstrated that a proper dispersion design profile significantly improves efficiency [15]. This causes the Stokes and anti-stokes bands generated by four-wave mixing (FWM) directly from the pump to broaden and merge, resulting in an 800 nm-wide SC source. Coen *et al* observed a similar combination of Raman scattering and FWM in an experiment in which a pulse that had a width of 60 ps and an energy of 40 nJ at 647 nm generated a 450 THz SC from 400 to 1000 nm in the fundamental mode using 10 m of PCF at the ZDW, i.e. 675 nm [16]. Aside from the femtosecond and picosecond regimes, a reasonable study of nanosecond pulses in SCG has also been conducted using PCFs. Dudley *et al* generated a SC from 460 to 750 nm with a bandwidth of over 250 THz in 1.8 m of PCF using 0.8 ns 300 nJ energy pulses from a Q-switched microchip laser at 532 nm, via excitation of a higher-order mode whose ZDW at 580 nm was reached from the pump wavelength via cascaded Raman scattering [17]. In addition, SCG has benefited from the development of high-power CW fiber sources. At 1065 nm, a Yb3+-doped fiber amplifier has been used in a master oscillator power fiber amplifier. Avodkhin *et al* used a 100 m-long PCF to generate a SC using high-power CW fiber sources in the 1065–1375 nm wavelength range [18]. Aside from pulse duration, the effects of the input pulse parameters such as pulse energy, peak power, and central wavelength on SCG in PCF have received a lot of attention and have been thoroughly investigated. Falk *et al* demonstrated that in tapered PCFs pumped with 15 fs pulses with an average power of a milliwatt, two distinct smooth spectral parts were generated, which had improved stability due to normal dispersion [19].

The fibers of choice have traditionally been silica PCFs and doped materials because they can be made in an infinite number of single-mode and adjustable ZDW configurations. SCG in non-silica technologies, using either nonlinear liquid fillings or different dopant materials, has recently piqued the interest of many researchers due to its enhanced nonlinearity. Zhang *et al* used liquid-core PCFs (LCPCFs) with CS_2 and nitrobenzene filling the core pumped at 1.55 μm by subpicosecond pulses to generate dramatically broadened SC in a range from 700 nm to more than 2500 nm [20]. Furthermore, SCG in chloroform-filled LCPCF has been demonstrated at 800 nm [21]. In addition, SCG with two-octave spectral coverage from 410 to 1640 nm has been experimentally demonstrated using a pump wavelength of 1200 nm and a few microjoule pump pulses in water-filled LCPCF [22]. Because of its higher nonlinearity and low transmission loss in the mid-infrared region at 2–4 μm, another type of PCF made of soft glass has gained popularity for mid-infrared SCG [23–26]. Soft glass fibers have a low loss in the mid-infrared region and nonlinearity that is up to 800 times stronger than that of silica. Fluoride, chalcogenide, and tellurite fibers, in particular, have been studied and demonstrated to have excellent optical properties for achieving a broad and flat SC spanning from the visible to the mid-infrared region. For example, using 110 fs pulses at 1550 nm, an SC with a bandwidth greater than 4 μm was generated in a short tellurite fiber. [23]. In a recent experiment, a centimeter-long zirconium, barium, lanthanum, aluminium and sodium (ZBLAN) fluoride fiber was pumped in the normal dispersion regime by a 1450 nm femtosecond laser, and despite having a nonlinearity comparable to that of silica, an ultrabroad SC spanning from the ultraviolet to $\approx$6 μm was demonstrated. [25]. Simultaneously, Ole Bang *et al* demonstrated the formation of SCG in tellurite PCFs specifically designed for high-power picosecond pumping at the thulium wavelength of 1930 nm. They were able to achieve a maximum bandwidth of 4.6 μm using the PCF with the smallest pitch at an optimum length of only 2.8 cm. [26].

In general, the aim of supercontinuum generation research has been to increase bandwidth. PCFs are not required to obtain broad supercontinua, but they are more mechanically stable and easier to mass-produce than fiber tapers. PCF supercontinua have been used in spectroscopy, optical coherence tomography, pulse compression, tunable lasers, and optical frequency metrology. Among the various nonlinear phenomena, a key nonlinear process known as the soliton is crucial in SCG-induced pulse compression.

9.2 Physical mechanisms

9.2.1 Mechanism of supercontinuum generation

The interplay of various nonlinear effects such as SPM, Raman scattering, and FWM can be used to interpret the detailed physical aspects of SCG (FWM). Solitons are nonlinear phenomena that serve as the foundation for the SCG mechanism. Husakou *et al* proposed a comprehensive theory of SCG based on soliton-related effects [26], which has been validated by experiments and numerical simulations in the femtosecond regime [27]. Among the soliton-related effects, the soliton frequency shift induced by Raman scattering and the emission of dispersive radiation due to third-order dispersion (TOD) play critical roles in SCG. We will consider the case of pumping in the anomalous dispersion regime, but close to the fiber's ZDW, as a conceptually clear way of understanding the main features of SCG in the femtosecond regime. The power of the pump pulse is high enough under typical pumping conditions for the input pulses to be considered as solitons of order N. A higher-order soliton pulse, N, disintegrates into N constituent red-shifted solitons with varying group velocities (figure 9.1). The fundamental soliton's energy does not change in principle, but it emits spectral components on the shorter wavelength (anti-Stokes) side of the pulse spectrum. Non-solitonic radiation (NSR), also known as dispersive radiation, exists in the presence of such blue-shifted radiation (figure 9.2). The wavelength of the NSR is determined by the phase-matching condition that results from third- and higher-order dispersion perturbation. The phases of non-solitonic radiation at ω_d and solitons at the frequency ω_s at a distance z after a delay $T = z/v_g$ are given by [28]

$$\phi_{\omega_d} = \beta(\omega_d)z - \omega_d(z/v_g) \tag{9.1}$$

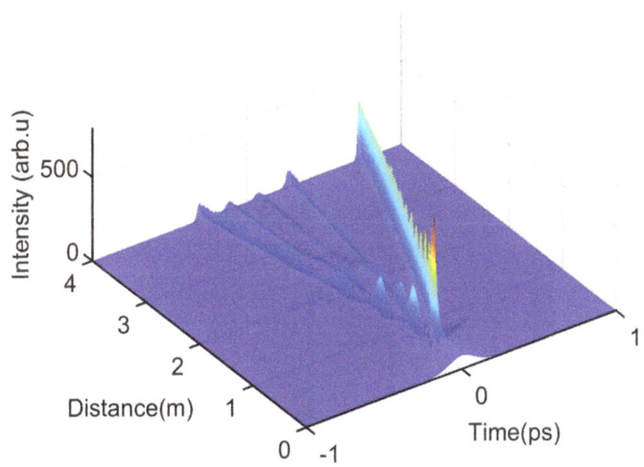

Figure 9.1. Evolution of a pulse produced by the fission of a higher-order soliton.

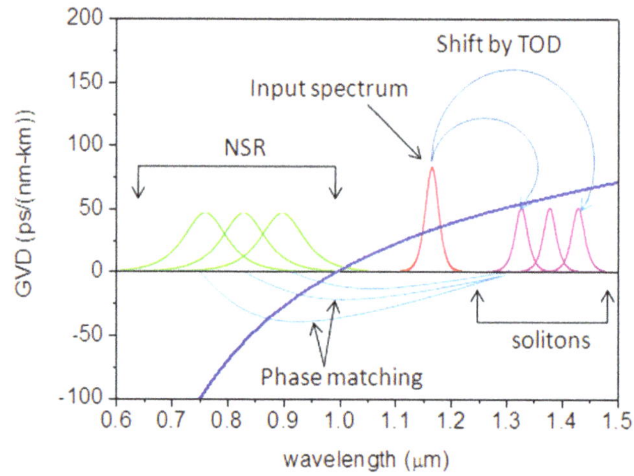

Figure 9.2. Schematic of SCG through the soliton fission process. The thick blue line represents the GVD of a PCF. The spectra of NSR, solitons, and the input pulse are represented by green, magenta, and red lines, respectively. The phase mismatch between solitons and NSR is represented by light blue lines.

$$\phi_{\omega_s} = \beta(\omega_s)z - \omega_s(z/v_g) + \frac{1}{2}\gamma P_s z, \tag{9.2}$$

where P_s denotes the peak power of the Raman soliton formed following the fission process. The final term in equation (9.2) represents the nonlinear phase shift caused by the soliton. As a result of a Raman-induced frequency shift, the soliton central frequency ω_s changes. The frequency of the dispersive wave changes as well. When we expand $\beta(\omega_d)$ in a Taylor series around ωs, the two phases match when the frequency shift $\Omega_d = \omega_d - \omega_s$ satisfies

$$\sum_{n=2}^{\infty} \frac{\beta_n(\omega_s)}{n!} \Omega_d^n = \frac{1}{2}\gamma P_s. \tag{9.3}$$

It is clear from equation (9.3) that there is no solution for Ω_d if the higher-order dispersive terms are absent and β_{20}. If we include the TOD, the resulting cubic polynomial, $\beta_3\Omega_d^3 + 3\beta_2\Omega_d^2 - 3\gamma P_s = 0$, provides the following approximate solution:

$$\Omega_d \approx -\frac{3\beta_2}{\beta_3} + \frac{\gamma P_s \beta_3}{3\beta_2^2}. \tag{9.4}$$

The frequency shift Ω_d is positive for soliton propagation in the anomalous dispersion regime, where $\beta_2 < 0$ and $\beta_3 > 0$. As a result, the NSR is emitted at a higher frequency (or a shorter wavelength) than that of the soliton. This situation changes in fibers with $\beta_3 < 0$, where the NSR can be emitted at wavelengths longer than that of the soliton. Higher-order dispersion effects are stronger in PCFs than in standard fibers and play a much larger role in pulse propagation. In the presence of

Raman effects and TOD, the generation of multiple frequency components due to soliton fission is much more pronounced in PCFs. The existence of multiple solitons with different frequencies results in a broad spectrum as a result of nonlinear interactions between the soliton and the blue-shifted continuum. In general, SCG is typically accomplished through two mechanisms: soliton fission and modulation instability (MI).

(i) Soliton fission

Soliton fission produces an ultrabroadband spectrum with pulse breaking caused primarily by the higher-order effects of soliton-related dynamics, such as higher-order linear dispersion and nonlinear Raman scattering. The physical phenomena of soliton fission and pulse spectral broadening in fiber have already been thoroughly investigated. For example, Dianov *et al* discovered the soliton Raman self-scattering effect in 1985 [29]. The authors reported Raman self-pumping of the Stokes-frequency spectral components of the same pulse by injecting a sufficiently powerful (capable of producing $N = 30$ solitons) input into a single-mode quartz optical fiber, as well as the generation of a broad soliton SC during stimulated Raman self-scattering of $N = 30$ wave packets. In 1986, Mitschke and Mollenauer reported an increasing red shift in the central frequency of a subpicosecond soliton pulse with increasing power in a standard single-mode, polarization-maintaining fiber [30]. Following the abovementioned theoretical prediction and experimental realization of broadband continua based on soliton effects, the latter part of the decade saw a thorough investigation of this mechanism for broadband generation using a variety of pump laser sources and pump durations.

(ii) Modulational-instability-induced supercontinuum generation

One of the most intriguing manifestations of MI is a method of obtaining an ultrabroadband spectrum [31, 32]. In this method, the FWM mechanism is in charge of a controlled MI process, which allows for manipulation and enhancement of the SCG process. Early investigations of modulational-instability-induced supercontinuum generation (MI-SCG) were carried out using conventional fiber in the low-dispersion regime to improve the broadband spectrum; later, the idea was successfully implemented in PCF. For example, it was demonstrated in 1983 that a continuum spectrum can be produced by the superposition of sequential stimulated Raman scattering (SRS) and FWM processes in a multimode fiber using two pump wavelengths by the reducing pump power [33]. Serkin also studied the structure of the field characteristics formed in the region of maximal self-compression of an N-soliton wave packet in a fiber in 1987. It was demonstrated that the noise component of the field causes bound-state decay in solitons, stochastic instability of the N-soliton pulse, and soliton noise generation. He investigated the possibility of an experimental realization of the decay of a soliton laser's output pulses into 'colored' envelope solitons, i.e. 'long-lived' (in comparison to the length scale of the dispersive spreading) nonlinear wave packets with a frequency shift [34]. When an auxiliary fiber with a chromatic-dispersion spectrum shifted to the long-wave direction is used, this possibility appears. Such soliton compression, fragmentation, and spectral shifting are key mechanisms in SCG's extension to long wavelengths.

9.3 Pulse compression through SCG

The generation of ultrashort pulses using SCG has recently piqued the interest of researchers [35–37]. If the phase of an SC spectrum can be completely synchronized, extremely effective short-pulse formation across the entire spectral domain is possible. Theoretically, the SCG process can be optimized to allow exact compensation of the SC phase shift by a liquid-crystal spatial light modulator. Several devices have been proposed for this purpose. For example, as reported in [38, 39] a 4.5 fs pulse was obtained in 1997 by passing an SC through an Ar-filled fiber, which emerged as an intense pulse after propagation through a pair of chirp-inducing mirrors. There is a practical cutoff limit in this process of achieving short pulses with broad spectral coverage. In particular, spectral coherence suffers in the case of broad spectra, despite being dependent on the chosen regime of fiber dispersion and the initial pulse duration. In theory, coherence is a property of the pulse that is extremely sensitive to noise in the initial pulses, resulting in beam-pointing instability and, ultimately, poor efficiency.

9.4 Tunable pulse compression

The effects of the input pulse parameters on SCG are a subject of great interest that has been thoroughly researched in silica-core PCFs for the last 15 years. The physics of SCG are now well understood, and the emphasis is shifting toward developing compact SCG sources for new applications and spectral windows [2, 40]. Among the numerous applications that make use of such broadband light sources, a few benefit from their tunable bandwidth property. Several techniques have been proposed in order to achieve tunable SC bandwidth. For example, in the study reported in [41], a tunable filter was used to obtain waveband tunability of the SC spectrum over a tunable range of 35 nm to 155 nm in an erbium-doped fiber ring cavity. In the study reported in [42], the pump wavelength was used to tune the bandwidth of a microstructured fiber's SC spectrum. The duration of the pump pulse has also been found to influence the bandwidth of the SC spectrum [43]. In this context, this chapter elaborates the numerical modeling of a PCF with the objective of obtaining a tunable SC spectrum. Since the refractive index of a medium is known to be affected by temperature, here, temperature is used as a control parameter in order to increase the tunable range of the SC's bandwidth [44–46]. The refractive-index changes that occur with temperature have an impact on a medium's optical properties. According to our findings, the ZDW of an optical fiber can be tuned by changing its temperature. Such variations in ZDW are known to affect the bandwidth of the SC spectrum [47, 48]. Based on these statements, the following sections will investigate the numerical modeling of PCF for SC generation and the influence of temperature on the tunability of the SC bandwidth.

9.5 Theoretical model

9.5.1 Fiber design

To investigate the effect of temperature on the dynamics of SCG-induced pulse compression, a PCF that has anomalous dispersion at 1.06 μm is numerically modeled. The PCF design procedure is as follows: the PCF's wavelength-dependent effective refractive indices are calculated using the fully vectorial effective index method (FVEIM) [49]. Silica's temperature-dependent refractive indices are calculated using Sellmeier's equation [50]:

$$
\begin{aligned}
n^2(\lambda, T) = {} & (1.315\,52 + 6.907\,54 \times 10^{-6}T) \\
& + \frac{(0.788\,404 + 23.5835 \times 10^{-6}T)\lambda^2}{\lambda^2 - (0.011\,0199 + 0.584\,758 \times 10^{-6}T)} \\
& + \frac{(0.913\,16 + 0.548\,368 \times 10^{-6}T)\lambda^2}{\lambda^2 - 100},
\end{aligned}
\tag{9.5}
$$

where T is the desired temperature for which refractive indices must be calculated. The dispersion parameters are estimated after the effective refractive indices have been calculated. According to numerical simulation, a PCF structure with an air hole diameter of 0.7 μm and a pitch value of 1 μm exhibits anomalous dispersion at 1.064 μm for temperature variations ranging from 100 °C to 1300 °C.

The refractive indices of silica and the effective refractive indices of the proposed PCF are plotted as a function of temperature in figure 9.3. It is discovered that all of the aforementioned parameters vary significantly as a function of temperature in a linear fashion. For example, as the temperature of the proposed PCF increases from 100°C to 1300°C, the refractive index of silica shifts from 1.4512 μm to 1.4635 μm. In figure 9.3, the coefficients of the effective refractive indices are also plotted for different temperatures. According to figure 9.3, the value of n_{eff} can be tuned from 1.378 to 1.3911 by changing the temperature from 100 °C to 1300 °C.

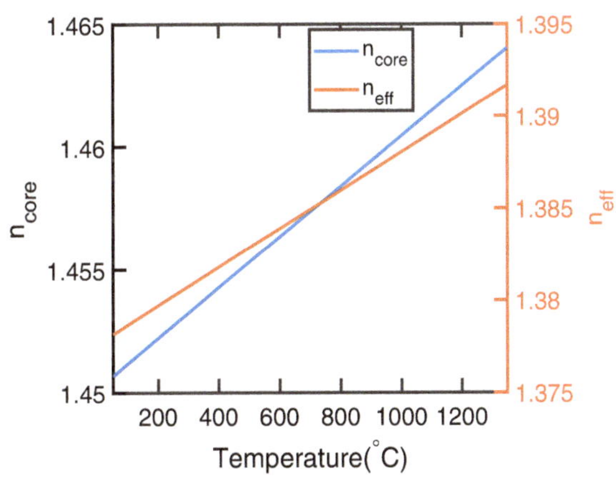

Figure 9.3. Variation of the refractive index of the core and the effective refractive index of the PCF as a function of temperature.

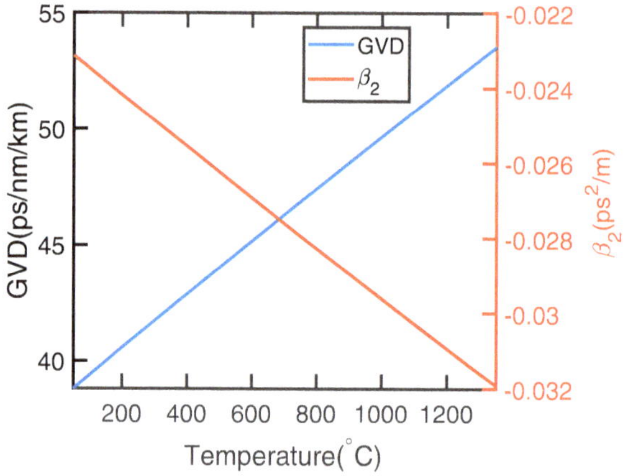

Figure 9.4. Variation of GVD and β_2 of the PCF as a function of temperature.

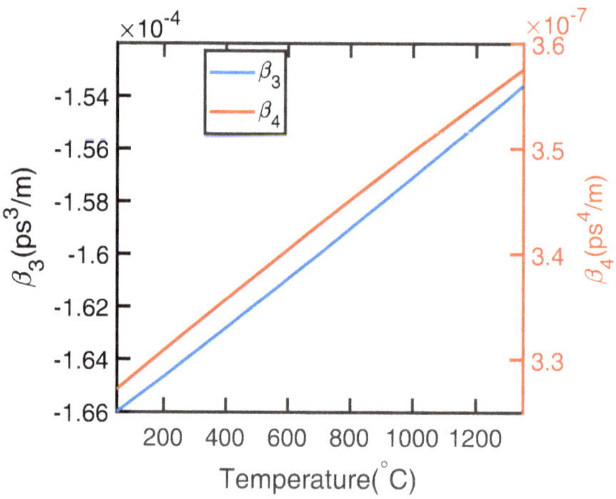

Figure 9.5. Variation of β_3 and β_4 as a function of temperature.

Figure 9.4 depicts the calculated fiber parameters β_2 and GVD with respect to the wavelength of the proposed PCF for various temperature values. All of the other fiber parameters change as the effective refractive indices change with temperature. As shown in figure 9.4, the GVD value shifts from 39.41 ps^{-1} nm^{-1} km^{-1} to 52.98 ps^{-1} nm^{-1} km^{-1} as the temperature is varied. Figure 9.4 depicts a change in β_2 from -2.349×10^{-2} ps^2 m^{-1} to -3.158×10^{-2} ps^2 m^{-1}. When the temperature is changed from 100 °C to 1300 °C, the anomalous dispersion regime produces 10^{-2} ps^2 m^{-1}. Likewise, β_3 is tuned from -1.655×10^{-4} ps^3 m^{-1} to -1.541×10^{-4} ps^4 m^{-1} and β_4 changes from 3.28×10^{-7} ps^4 m^{-1} to 3.41×10^{-7} ps^4 m^{-1}, as illustrated in figure 9.5. As shown in figure 9.6, the variation in nonlinearity clearly indicates that the temperature can change its values. Thus, the proposed PCF's dispersion parameters and nonlinearity are found to be temperature tunable.

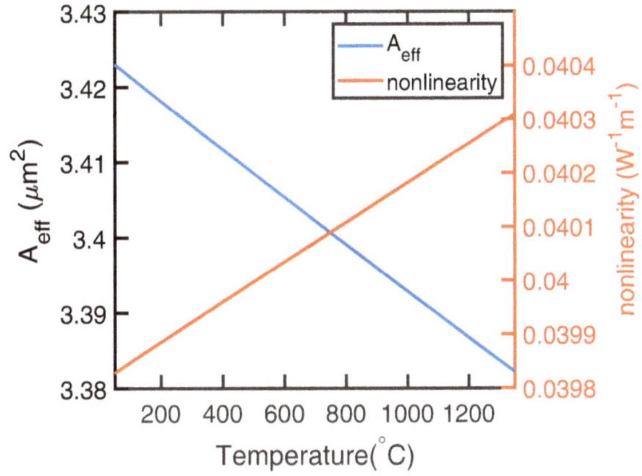

Figure 9.6. Variation of the effective core area and the nonlinearity of the PCF as a function of temperature.

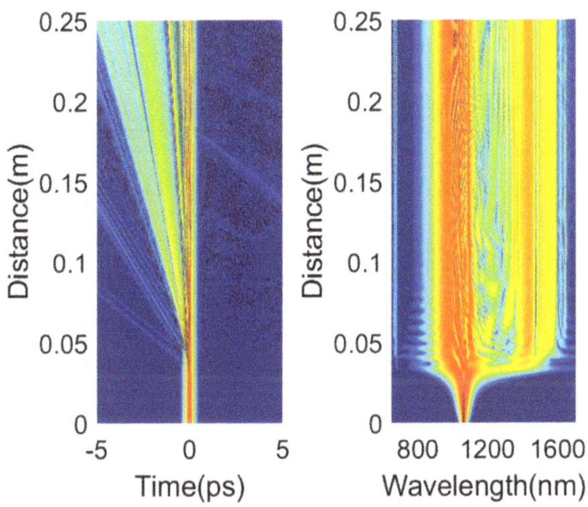

Figure 9.7. Temporal and spectral profiles of pulse propagation in PCF for 100 °C.

In order to analyze the effect of temperature on pulse compression in the proposed PCF, it is best to first understand the dynamics of the SCG pulse. As a result, the discussion begins with a look at the propagation of a hyperbolic secant pulse with a width of 140 fs and a peak power of 4.5 kW, which corresponds to a 7th-order single soliton in the proposed PCF. The spectral and temporal profiles of the hyperbolic secant pulse at a temperature of 100 °C are depicted in figure 9.7; the pulse is launched at 1064 nm and falls in the anomalous dispersion region of the proposed water core PCF. The input pulse is seen to split into several solitons, which manifest as stable peaks in the temporal shape. From this, we conclude that the solitons remain stable even when disturbed by slow nonlinearity and the presence of other solitons. After fission, the fast component dominates the dynamics of the shortest sub-100 fs solitons. The spectral broadening in the presence of Raman

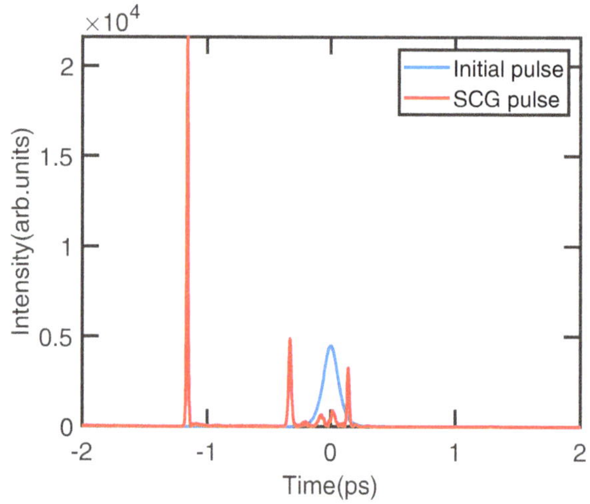

Figure 9.8. Compressed pulse at the end of PCF.

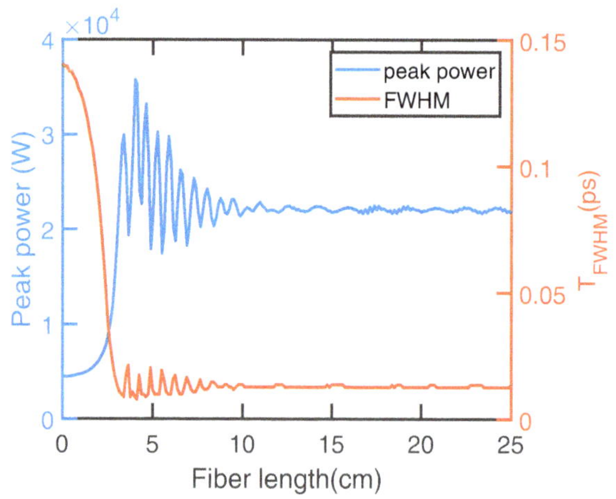

Figure 9.9. Variation of peak power and pulse width along the fiber length.

response demonstrates that the Raman response has a significant impact on spectral broadening. When the Raman response is included, the spectrum of a higher-order soliton, the generation of a distinct peak of non-solitonic radiation by the soliton frequency shift, and a one-octave-broad spectrum are predicted for the aforementioned fiber parameters. The final compressed pulse profile is shown in figure 9.8. The pulse width of the input pulse decreases significantly from 140 fs to 11.72 fs as the dynamics of SC pulses change, as shown in figure 9.9. Despite the fact that efficient compression with a compression factor of 11.9 has been achieved in a fiber length of 25 cm using this SCG compression scheme, simulations show that 53% of the input energy takes the form of pedestal energy for this compression scheme.

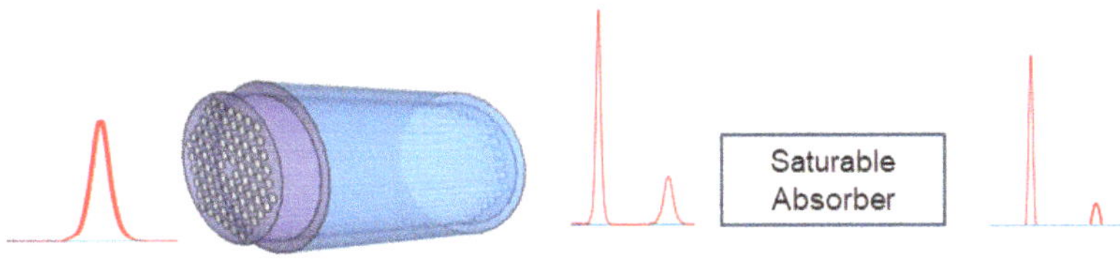

Figure 9.10. Schematic representation of the SCG-induced pulse-compression scheme.

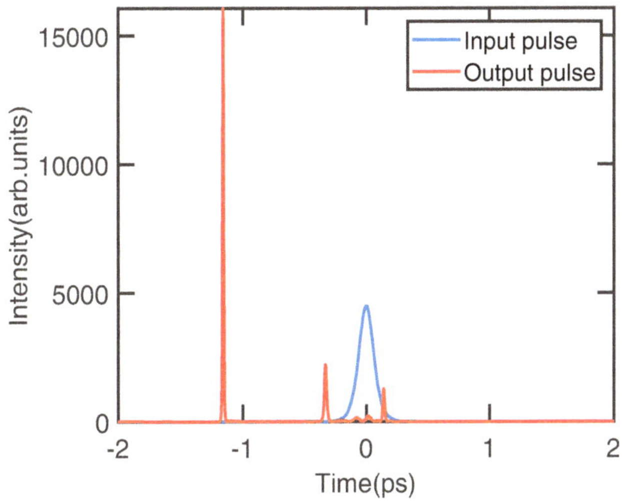

Figure 9.11. Compressed pulse at the end of the saturable absorber.

To overcome this drawback, the fiber compressor structure was modified to include a saturable absorber after the previously proposed PCF, as shown in figure 9.10. The power transfer function of the saturable absorber can be found from the following:

$$T = 1 - \frac{\alpha_{ns}}{1 + \dfrac{P}{P_{sat}}}, \tag{9.6}$$

where α_{ns} and P_{sat} are the nonsaturated loss and the saturation power, respectively; values of $\alpha_{ns} = 0.81$ and $P_{sat} = 10$ kW were used in the simulation. We can now analyze the dynamics of the SC pulse in the saturable absorber, which suppresses the low-power components of the SC pulse as plotted in figure 9.8. Thus, it can be observed that one can compress the 140 fs width of the pulse to 11.65 fs with an output pedestal energy of 3.85% through an SCG-induced pulse-compression scheme (figure 9.11).

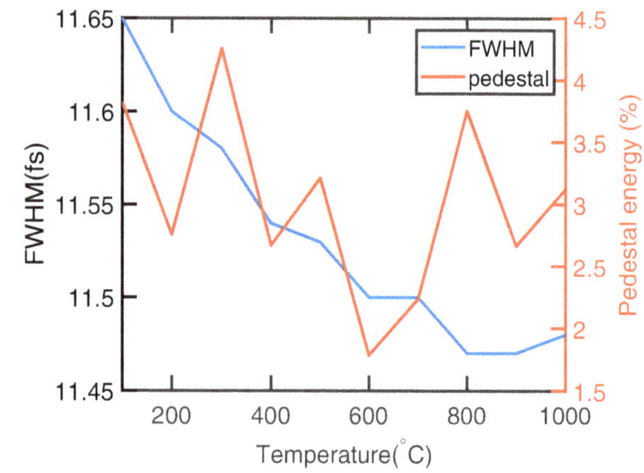

Figure 9.12. Relation between temperature, pulse width, and pedestal energy of SC-induced compression.

9.5.2 Temperature-dependent pulse compression

In order to quantify the variation in bandwidth caused by the change in external temperature surrounding the fiber and the width of the propagating pulse, the experimental conditions are incorporated into the theoretical analysis by taking an ensemble average of a set of realizations of the spectral profile at the PCF's output end. As a result, a sufficient number of simulations were conducted to investigate the shot-to-shot variation of the SC, in which the initial conditions of each shot were slightly modified to contain additional noise. The averaged output spectrum of all shots is shown in figure 9.12, along with the final pulse width and pedestal energy as a function of temperature. One can also notice a perturbation in the pulse's trailing edge, although it has a stable leading edge. Further inspection of the plot shows that the pulse parameters such as peak intensity and spectral center also change as a function of temperature. The perturbation, which takes the form of a time delay in the higher-wavelength region, increases as the temperature decreases. This kind of perturbation, which occurs in the trailing edge of the pulse, leads to broadening in the longer-wavelength region of the spectrum. For femtosecond pulse propagation controlled by temperature, figure 9.12 shows that the width of the generated SC-induced pulse compression decreases as the temperature increases.

References

[1] Zheltikov A M 2006 Let there be white light: supercontinuum generation by ultrashort laser pulses *Phys.-Usp.* **49** 605

[2] Dudley J M, Genty G and Coen S 2006 Supercontinuum generation in photonic crystal fiber *Rev. Mod. Phys.* **78** 1135

[3] Dudley J M and Taylor J R 2010 *Introduction and history Supercontinuum Generation in Optical Fibers* (Cambridge: Cambridge University Press) Introduction and history

[4] Skryabin D V and Gorbach A V 2010 Colloquium: looking at a soliton through the prism of optical supercontinuum *Rev. Mod. Phys.* **82** 1287

[5] Alfano R R and Shapiro S L 1970 Emission in the region 4000 to 7000 å via four-photon coupling in glass *Phys. Rev. Lett.* **24** 584
[6] Lin C and Stolen R H 1976 New nanosecond continuum for excited-state spectroscopy *Appl. Phys. Lett.* **28** 216–8
[7] Beaud P, Hodel W, Zysset B and Weber H 1987 Ultrashort pulse propagation, pulse breakup, and fundamental soliton formation in a single-mode optical fiber *IEEE J. Quantum Electron.* **23** 1938–46
[8] Morioka T, Mori K and Saruwatari M 1993 More than 100-wavelength-channel picosecond optical pulse generation from single laser source using supercontinuum in optical fibres *Electron. Lett.* **29** 862–4
[9] Takushima Y, Futami F and Kikuchi K 1998 Generation of over 140-nm-wide supercontinuum from a normal dispersion fiber by using a mode-locked semiconductor laser source *IEEE Photonics Technol. Lett.* **10** 1560–2
[10] Morioka T, Uchiyama K, Kawanishi S, Suzuki S and Saruwatari M 1995 Multiwavelength picosecond pulse source with low jitter and high optical frequency stability based on 200 nm supercontinuum filtering *Electron. Lett.* **31** 1064–6
[11] Ranka J K, Windeler R S and Stentz A J 2000 Visible continuum generation in air-silica microstructure optical fibers with anomalous dispersion at 800 nm *Opt. Lett.* **25** 25–7
[12] Roy S, Ghosh D, Bhadra S K and Agrawal G P 2010 Role of dispersion profile in controlling emission of dispersive waves by solitons in supercontinuum generation *Opt. Commun.* **283** 3081–8
[13] Price J H V, Furusawa K, Monro T M, Lefort L and Richardson D J 2002 Tunable, femtosecond pulse source operating in the range 1.06–1.33 μm based on an Yb 3+-doped holey fiber amplifier *J. Opt. Soc. Am.* B **19** 1286–94
[14] Nishizawa N, Chen Y, Hsiung P, Ippen E P and Fujimoto J G 2004 Real-time, ultrahigh-resolution, optical coherence tomography with an all-fiber, femtosecond fiber laser continuum at 1.5 μm *Opt. Lett.* **29** 2846–8
[15] Nikolov N I, Sørensen T, Bang O and Bjarklev A 2003 Improving efficiency of supercontinuum generation in photonic crystal fibers by direct degenerate four-wave mixing *J. Opt. Soc. Am.* **B 20** 2329–37
[16] Coen S, Chau A H L, Leonhardt R, Harvey J D, Knight J C, Wadsworth W J and Russell P S J 2002 Supercontinuum generation by stimulated Raman scattering and parametric four-wave mixing in photonic crystal fibers *J. Opt. Soc. Am.* B **19** 753–64
[17] Dudley J M, Provino L, Grossard N, Maillotte H, Windeler R S, Eggleton B J and Coen S 2002 Supercontinuum generation in air-silica microstructured fibers with nanosecond and femtosecond pulse pumping *J. Opt. Soc. Am.* B **19** 765–71
[18] Avdokhin A V, Popov S V and Taylor J R 2003 Continuous-wave, high-power, Raman continuum generation in holey fibers *Opt. Lett.* **28** 1353–5
[19] Falk P, Frosz M H and Bang O 2005 Supercontinuum generation in a photonic crystal fiber with two zero-dispersion wavelengths tapered to normal dispersion at all wavelengths *Opt. Express* **13** 7535–40
[20] Zhang R, Teipel J and Giessen H 2006 Theoretical design of a liquid-core photonic crystal fiber for supercontinuum generation *Opt. Express* **14** 6800–12
[21] Zhang H, Chang S, Yuan J and Huang D 2010 Supercontinuum generation in chloroform-filled photonic crystal fibers *Optik* **121** 783–7

[22] Bethge J, Husakou A, Mitschke F, Noack F, Griebner U, Steinmeyer G and Herrmann J 2010 Two-octave supercontinuum generation in a water-filled photonic crystal fiber *Opt. Express* **18** 6230–40

[23] Domachuk P, Wolchover N A, Cronin-Golomb M, Wang A, George A K, Cordeiro C M B, Knight J C and Omenetto F G 2008 Over 4000 nm bandwidth of mid-IR supercontinuum generation in sub-centimeter segments of highly nonlinear tellurite PCFs *Opt. Express* **16** 7161–8

[24] Kanth Kumar V V R, George A K, Knight J C and Russell P S J 2003 Tellurite photonic crystal fiber *Opt. Express* **11** 2641–5

[25] Qin G, Yan X, Kito C, Liao M, Chaudhari C, Suzuki T and Ohishi Y 2009 Ultrabroadband supercontinuum generation from ultraviolet to 6.28 μm in a fluoride fiber *Appl. Phys. Lett.* **95** 161103

[26] Buccoliero D, Steffensen H, Bang O, Ebendorff-Heidepriem H and Monro T M 2010 Thulium pumped high power supercontinuum in loss-determined optimum lengths of tellurite photonic crystal fiber *Appl. Phys. Lett.* **97** 061106

[27] Herrmann J, Griebner U, Zhavoronkov N, Husakou A, Nickel D, Knight J C, Wadsworth W J, Russell P S J and Korn G 2002 Experimental evidence for supercontinuum generation by fission of higher-order solitons in photonic fibers *Phys. Rev. Lett.* **88** 173901

[28] Agrawal G P 2007 *Nonlinear Fiber Optics* 4 (Cambridge, MA: Academic Press)

[29] Dianov E M and Mamyshev V *et al* 1985 Stimulated-Raman conversion of multisoliton pulses in quartz optical fibers *JETP Lett.* **41** 242–244

[30] Mitschke F M and Mollenauer L F 1986 Discovery of the soliton self-frequency shift *Opt. Lett.* **11** 659–61

[31] Hickmann J M, Cavalcanti S B, Borges N M, Gouveia E A and Gouveia-Neto A S 1993 Modulational instability in semiconductor-doped glass fibers with saturable nonlinearity *Opt. Lett.* **18** 182–4

[32] Demircan A and Bandelow U 2005 Supercontinuum generation by the modulation instability *Opt. Commun.* **244** 181–5

[33] Nakazawa M and Tokuda M 1983 Continuum spectrum generation in a multimode fiber using two pump beams at 1.3 μm wavelength region *Japan J. Appl. Phys.* **22** L239

[34] Serkin V N 1987 Colored envelope solitons in optical fibers *Sov. Tech. Phys. Lett.* **13** 320–1

[35] Schenkel B, Paschotta R and Keller U 2005 Pulse compression with supercontinuum generation in microstructure fibers *J. Opt. Soc. Am.* B **22** 687–93

[36] Dudley J M and Coen S 2004 Fundamental limits to few-cycle pulse generation from compression of supercontinuum spectra generated in photonic crystal fiber *Opt. Express* **12** 2423–8

[37] Joshva Raj G, Raja R V J, Grelu P, Ganapathy R and Porsezian K 2015 Effect of temperature on supercontinuum generation in water-core photonic crystal fiber *IEEE Photonics Technol. Lett.* **28** 1209–12

[38] Spence D E, Kean P Np and Sibbett W 1991 60 -fsec pulse generation from a self-mode-locked Ti: sapphire laser *Opt. Lett.* **16** 42–4

[39] Nisoli M, De Silvestri S, Svelto O, Szipöcs R, Ferencz K, Spielmann Ch, Sartania S and Krausz F 1997 Compression of high-energy laser pulses below 5 fs *Opt. Lett.* **22** 522–4

[40] Dudley J M and Taylor J R 2010 *Supercontinuum Generation in Optical Fibers* (Cambridge: Cambridge University Press)

[41] Lin H and Wang Y-W 2014 Waveband-tunable supercontinuum fiber laser *Opt. Commun.* **333** 22–5
[42] Kobtsev S M, Kukarin S V and Smirnov S V 2008 Fiber supercontinuum generator with wavelength-tunable pumping *Laser Phys.* **18** 1257–9
[43] Andreana M *et al* 2012 Control of near-infrared supercontinuum bandwidth by adjusting pump pulse duration *Opt. Express* **20** 10750–60
[44] George Falk K 1909 The change in refractive index with temperature. I *J. Am. Chem. Soc.* **31** 86–107
[45] George Falk K 1909 The change in refractive index with temperature II *J. Am. Chem. Soc.* **31** 806–21
[46] Nallusamy N, Raja R V J and Joshva Raj G 2017 Highly sensitive nonlinear temperature sensor based on modulational instability technique in liquid infiltrated photonic crystal fiber *IEEE Sens. J.* **17** 3720–7
[47] Hansryd J, Dross F, Westlund M, Andrekson P A and Knudsen S N 2001 Increase of the SBs threshold in a short highly nonlinear fiber by applying a temperature distribution *J. Lightwave Technol.* **19** 1691
[48] Joshva Raj G, Raja R V J, Nagarajan N and Ramanathan G 2016 Tunable broadband spectrum under the influence of temperature in IR region using cs_2 core photonic crystal fiber *J. Lightwave Technol.* **34** 3503–9
[49] Raja R V J and Porsezian K 2007 A fully vectorial effective index method to analyse the propagation properties of microstructured fiber *Photonics Nanostruct. Fundam. Appl* **5** 171–7
[50] Yamunadevi R, Raja R V J and Arzate N 2018 Generation of low power and ultrashort laser pulses at 800 nm through soliton compression in chloroform-infiltrated cascaded photonic crystal fibers *J. Appl. Phys.* **124** 113105

Appendix A

MATLAB®

Determination of FWHM

```
function width = fwhm1(x,t)
[mm nn] = size(x);

xx=abs(x).^2;
peak=max(xx);
ind=find(xx==peak);

half_peak = peak(1)/2 ;
for iii=1:nn-1
    xx(iii);
if (xx(iii)<=half_peak) && (xx(iii+1)>half_peak)
break;
end
end
t(iii);
for jjj=ind:nn-1
    jjj;
if (xx(jjj)>=half_peak) && (xx(jjj+1)<half_peak)
break;
end
end

t(jjj);
width = t(jjj)-t(iii);
```

doi:10.1088/978-0-7503-2686-5ch10

Appendix B

MATLAB®

doi:10.1088/978-0-7503-2686-5ch11

Higher-order soliton compression

```
clear all
close all
format long

%=============================================
% Physical Parameters
%=============================================
c=3*10^8;                          %[m/sec] speed of light
lamda=0.8e-6;                     %[m] center wavelength
omeg_0=2*pi*c/lamda;                  % center angular
frequency
f_rep = 80e6;                      % Pulse repetition rate
[Hz]

%=============================================
% fiber Parameters
%=============================================
betas = [-1.0483046398 0.000037889343 0.000001020432];
beta2= betas(1);   % second order dispersion ps^2/m
beta3=betas(2);     % third order dispersion ps^3/m
beta4=betas(3);     % fourth order dispersion ps^4/m
gamma=0.15;         % Nonlinear parameter [(W*m)^-1]
loss=0.00000045;                        % [dB/m] Power Loss
alpha=log(10.^(loss/10));       %[1/m] loss

%=============================================
% pulse Parameters
%=============================================
fwhm=10;                          % [ps] full width half maximum
pulsewidth=fwhm/(2*log(1+sqrt(2)));
%=============================================
% Time Window and Frequency Window
%=============================================
T_0 = pulsewidth*20;                    % size of time window
N_t=2^12;                           % The number of points
dt=T_0/N_t;                          % Temporal resolution
[ps]
t=linspace(-dt*N_t/2, dt*(N_t/2-1), N_t); % time window
```

```
dw=2*pi/(N_t*dt);
w=linspace(-dw*N_t/2, dw*(N_t/2-1), N_t); % frequency
window
lams=2*pi*c.*1e9./(omeg_0+w*1e12) ;

%===============================================
% Input Signal
%===============================================

soliton_order =3;
P0=soliton_order^2*abs(beta2)/(gamma*pulsewidth^2);     %
input power [W]
chirp = 0; % Chirping Parameter
U0=sqrt(P0)*sech(t./pulsewidth);                      % input
pulse
V0 = fftshift(ifft(U0));
Q0 = V0.*conj(V0);

%===============================================
% Define Simulation Distance and Step Size
%===============================================
L_D=pulsewidth^2/abs(beta2);             % Dispersion length
Length_z=0.33*L_D;                       % Fiber length [m]
n_steps=1000;                     % Number of propagation
steps
dz = Length_z/n_steps;


N_saves = 40;
U_saved = zeros(N_t, N_saves+1);
V_saved = zeros(N_t, N_saves+1);
z_saved = linspace(0, Length_z, N_saves+1);

%===============================================
% Initialize input pulse
%===============================================
z=0;
U=U0;
Lin_opt = zeros(1,N_t);
for m = 2:length(betas)+1
   Lin_opt = Lin_opt + i * betas(m-1)/factorial(m) * w.^m;
end
disper = exp(Lin_opt * dz);
```

```
disper_half1 = exp(Lin_opt * dz/2);
disper_half2 = exp(Lin_opt * -dz/2);


U_saved(:,1)=U0;
V_saved(:,1)=V0;

nz_save = n_steps / N_saves;
inc=2;

%===============================================%
% Propagation Through Fiber %
%===============================================%
V = fftshift(ifft(U));
V = V.*disper_half1;
U = fft(fftshift(V));
for nz=1:n_steps
    n_steps-nz;
    P_present = U.*conj(U);
    U = U.*exp(dz*i*gamma*(P_present));
    V = fftshift(ifft(U));
    V = V.*disper;
    U = fft(fftshift(V));

    z = z + dz;
    if (nz_save == nz)

        V_present = V.*disper_half2;
        U_present = fft(fftshift(V_present));
        figure(111)
        subplot(1,2,1);

plot3(t(1:10:end),z*ones(size(t(1:10:end))),abs(U_present(1
:10:end)).^2,'-','Markersize',2);
            grid on
            hold on
        subplot(1,2,2);

plot3(lams(1:10:end),z*ones(size(lams(1:10:end))),abs(V_pre
sent(1:10:end)).^2,'-','Markersize',2);
            grid on
            hold on
        drawnow;
          if (nz_save == n_steps)
```

```
            U_compressed=U_present;
        end

        U_saved(:, inc)=U_present;
        V_saved(:, inc)=V_present;
        nz_save = nz_save + n_steps / N_saves;
        n_steps-nz_save
        inc=inc+1;
    end
end

V_present = V.*disper_half2;
U_present = fft(fftshift(V));
U_saved(:,end)=U_present;
V_saved(:,end)=V_present;

figure(1)
F1=plot(t,U0.*conj(U0),t,U_compressed.*conj(U_compressed),'
r');
set(F1,'linewidth',2);
set(gca,'linewidth',2,'fontsize',16,'Ticklength',[0.03
0.03]);
axis([-50 50 0 max(U_compressed.*conj(U_compressed))])
xlabel('Time(ps)','fontsize',16);
ylabel('Intensity(arb.units)','fontsize',16);


U_saved=U_saved';
V_saved=V_saved';
z_saved=z_saved';

figure(2)
hh2=mesh(t,z_saved,U_saved.*conj(U_saved));
set(hh2, 'FaceColor', 'flat');
set(gca,'linewidth',2,'fontsize',18,'Ticklength',[0.03
0.03]);
shading interp;
set(hh2, 'FaceAlpha', 0.7);
xlabel('Time(ps)','fontsize',18);
ylabel('Distance(m)','fontsize',18);
zlabel('Intensity (arb.units)','fontsize',18)
```

```
figure(4);
inten_ln=10*log10(U_saved.*conj(U_saved));
max_int = max(max(inten_ln));
subplot(1,2,1);
pcolor(t, z_saved,inten_ln);
set(0,'DefaultFigureColormap',feval('jet'));
caxis([max_int-40.0, max_int]);  xlim([-50,50]); shading
interp;
set(gca,'linewidth',2,'fontsize',18,'Ticklength',[0.03
0.03]);
xlabel('Time(ps)','fontsize',18);
ylabel('Distance(m)','fontsize',18);

intfe_ln=10*log10(V_saved.*conj(V_saved));
max_inf = max(max(intfe_ln));
subplot(1,2,2);
pcolor(lams,z_saved,intfe_ln);
set(0,'DefaultFigureColormap',feval('jet'));
caxis([max_inf-40.0, max_inf]);  xlim([795,805]); shading
interp;
set(gca,'linewidth',2,'fontsize',18,'Ticklength',[0.03
0.03]);
xlabel('Wavelength(nm)','fontsize',18);
ylabel('Distance(m)','fontsize',18);


figure(5)
waterfall(t,z_saved,U_saved.*conj(U_saved));
set(gca,'linewidth',2,'fontsize',16,'Ticklength',[0.03
0.03]);
xlabel('Time(ps)','fontsize',18);
ylabel('Distance(m)','fontsize',18);
zlabel('Intensity (arb.units)','fontsize',18)
```

Appendix C

MATLAB®

doi:10.1088/978-0-7503-2686-5ch12

Adiabatic compression

```
clear all
close all
format long

%================================================
% Physical Parameters
%================================================
c=3*10^8;                          %[m/sec] speed of light
lamda=0.8e-6;                     %[m] center wavelength
omeg_0=2*pi*c/lamda;                  % center angular
frequency
f_rep = 80e6;                       % Pulse repetition rate
[Hz]

%================================================
% fiber Parameters
%================================================
betas = [-0.091333197019  0  0];
beta2= betas(1);   % second order dispersion ps^2/m
beta3=betas(2);     % third order dispersion ps^3/m
beta4=betas(3);     % fourth order dispersion ps^4/m
gamma=0.029668995110;          % initial Nonlinear
parameter [(W*m)^-1]
loss=0;                        % [dB/m] Power Loss
alpha=0;
bet_end=-0.040728146510;                % second order
dispersion at the end
gam_end=0.038164611659;              % Nonlinear parameter
at he end

%================================================
% pulse Parameters
%================================================
fwhm=10;                       % [ps] full width half maximum
pulsewidth=fwhm/(2*log(1+sqrt(2)));
%================================================
% Time Window and Frequency Window
%================================================
T_0 = pulsewidth*20;                     % size of time window
N_t=2^12;                           % The number of points
```

```
dt=T_0/N_t;                          % Temporal resolution
[ps]
t=linspace(-dt*N_t/2, dt*(N_t/2-1), N_t); % time window


dw=2*pi/(N_t*dt);
w=linspace(-dw*N_t/2, dw*(N_t/2-1), N_t); % frequency
window
lams=2*pi*c.*1e9./(omeg_0+w*1e12) ;

%===============================================
% Input Signal
%===============================================

soliton_order =1;
P0=soliton_order^2*abs(beta2)/(gamma*pulsewidth^2);     %
input power [W]
chirp = 0; % Chirping Parameter
U0=sqrt(P0)*sech(t./pulsewidth);                   % input
pulse
V0 = fftshift(ifft(U0));
Q0 = V0.*conj(V0);


%===============================================
% Define Simulation Distance and Step Size
%===============================================

L_D=pulsewidth^2/abs(beta2);            % Dispersion length
Length_z=4*L_D;                       % Fiber length [m]
n_steps=1000;                       % Number of propagation
steps
dz = Length_z/n_steps;

N_saves = 20;
U_saved = zeros(N_t, N_saves+1);
V_saved = zeros(N_t, N_saves+1);
z_saved = linspace(0, Length_z, N_saves+1);


dbet=(beta2-bet_end)/(n_steps-1);
dgam=(gamma-gam_end)/(n_steps-1);
```

```
%==============================================
% Initialize input pulse
%==============================================
z=0;
beta2_z=beta2;
gamma_z=gamma;
beta3_z=beta3;
beta4_z=beta4;

U=U0;
Lin_opt = zeros(1,N_t);
for m = 2:length(betas)+1
   Lin_opt = Lin_opt + i * betas(m-1)/factorial(m) * w.^m;
end
disper = exp(Lin_opt * dz);
disper_half1 = exp(Lin_opt * dz/2);
disper_half2 = exp(Lin_opt * -dz/2);


U_saved(:,1)=U0;
V_saved(:,1)=V0;

nz_save = n_steps / N_saves;


fwhm_pres=fwhm1(U,t);
pulsewidth_pres= fwhm_pres/(2*log(1+sqrt(2)));
peak_pres=max(abs(U).^2);
Energy_out=trapz(t,(U.*conj(U)));
Energy_sech=(2*peak_pres*fwhm_pres)/(2*log(1+sqrt(2)));

len_pres(1)=z;
F_c(1)=fwhm1(U0,t)/fwhm_pres;
PE(1)=(abs(Energy_out-Energy_sech)*100/Energy_out);

t_fwhm(1)=fwhm_pres;
p_p(1)=peak_pres;

inc=2;

%==================================================%
% Propagation Through Fiber %
%==================================================%
```

```
for nz=1:n_steps
    n_steps-nz;


    betas_DDF= [beta2_z  beta3_z   beta4_z];

    Lin_DDF_opt = zeros(1,N_t);
    for m = 2:length(betas_DDF)+1
        Lin_DDF_opt= Lin_DDF_opt + i * betas_DDF(m-
1)/factorial(m)*w.^m;
    end
    disper_DDF = exp(Lin_DDF_opt * dz);
    disper_DDF_half1 = exp(Lin_DDF_opt * dz/2);
    disper_DDF_half2 = exp(Lin_DDF_opt * -dz/2);

    P_present = U.*conj(U);
    U = U.*exp(dz*i*gamma*(P_present));
    V = fftshift(ifft(U));
    V = V.*disper_DDF;
    U = fft(fftshift(V));

    z = z + dz;

    beta2_z=beta2_z-dbet;
    gamma_z=gamma_z-dgam;

    if (nz_save == nz)
        V_present = V.*disper_DDF_half2;
        U_present = fft(fftshift(V_present));
        figure(111)
        subplot(1,2,1);

plot3(t(1:10:end),z*ones(size(t(1:10:end))),abs(U_present(1
:10:end)).^2,'-','Markersize',2);
            grid on
            hold on
        subplot(1,2,2);

plot3(lams(1:10:end),z*ones(size(lams(1:10:end))),abs(V_pre
sent(1:10:end)).^2,'-','Markersize',2);
            grid on
            hold on
        drawnow;
```

```
        fwhm_pres=fwhm1(U_present,t);
        pulsewidth_pres= fwhm_pres/(2*log(1+sqrt(2)));
        peak_pres=max(abs(U_present).^2);
        Energy_out=trapz(t,(U_present.*conj(U_present)));

Energy_sech=(2*peak_pres*fwhm_pres)/(2*log(1+sqrt(2)));

        len_pres(inc)=z;
        F_c(inc)=fwhm1(U0,t)/fwhm_pres;
        PE(inc)=(abs(Energy_out-
Energy_sech)*100/Energy_out);

        t_fwhm(inc)=fwhm_pres;
        p_p(inc)=peak_pres;

        U_saved(:, inc)=U_present;
        V_saved(:, inc)=V_present;
        nz_save = nz_save + n_steps / N_saves;
        n_steps-nz_save
        inc=inc+1;
    end

end % for indice

figure(1)
F1=plot(t,U0.*conj(U0),t,U_present.*conj(U_present),'r');
set(F1,'linewidth',2);
set(gca,'linewidth',2,'fontsize',16,'Ticklength',[0.03
0.03]);
axis([-50 50 0 max(U_present.*conj(U_present))])
xlabel('Time(ps)','fontsize',16);
ylabel('Intensity(arb.units)','fontsize',16);


figure(121)
yyaxis left
F1=plot(len_pres,p_p)
set(F1,'linewidth',2);
set(gca,'linewidth',2,'fontsize',16,'Ticklength',[0.03
0.03]);
xlabel('Fiber length (m)','fontsize',16);
ylabel('Peak power (W)','fontsize',16);

yyaxis right
```

```
F2=plot(len_pres,t_fwhm)
set(F2,'linewidth',2);
set(gca,'linewidth',2,'fontsize',16,'Ticklength',[0.03
0.03]);
xlabel('Fiber length(m)','fontsize',16);
ylabel('T_{FWHM}(ps)','fontsize',16);

figure(131)
F1=plot(len_pres,F_c)
set(F1,'linewidth',2);
set(gca,'linewidth',2,'fontsize',16,'Ticklength',[0.03
0.03]);
xlabel('Fiber length (m)','fontsize',16);
ylabel('Compression factor','fontsize',16);

yyaxis right
F2=plot(len_pres,PE)
set(F2,'linewidth',2);
set(gca,'linewidth',2,'fontsize',16,'Ticklength',[0.03
0.03]);
xlabel('Fiber length (m)','fontsize',16);
ylabel('Pedestal energy (%)','fontsize',16);

U_saved=U_saved';
V_saved=V_saved';
z_saved=z_saved';


figure(2)
hh2=mesh(t,z_saved,U_saved.*conj(U_saved));
set(hh2, 'FaceColor', 'flat');
set(gca,'linewidth',2,'fontsize',18,'Ticklength',[0.03
0.03]);
shading interp;
set(hh2, 'FaceAlpha', 0.7);
xlabel('Time(ps)','fontsize',18);
ylabel('Distance(m)','fontsize',18);
zlabel('Intensity (arb.units)','fontsize',18)


figure(4);
inten_ln=10*log10(U_saved.*conj(U_saved));
max_int = max(max(inten_ln));
subplot(1,2,1);
```

```
pcolor(t, z_saved,inten_ln);
set(0,'DefaultFigureColormap',feval('jet'));
caxis([max_int-40.0, max_int]);  xlim([-50,50]); shading
interp;
set(gca,'linewidth',2,'fontsize',18,'Ticklength',[0.03
0.03]);
xlabel('Time(ps)','fontsize',18);
ylabel('Distance(m)','fontsize',18);

intfe_ln=10*log10(V_saved.*conj(V_saved));
max_inf = max(max(intfe_ln));
subplot(1,2,2);
pcolor(lams,z_saved,intfe_ln);
set(0,'DefaultFigureColormap',feval('jet'));
caxis([max_inf-40.0, max_inf]);  xlim([795 805]); shading
interp;
set(gca,'linewidth',2,'fontsize',18,'Ticklength',[0.03
0.03]);
xlabel('Wavelength(nm)','fontsize',18);
ylabel('Distance(m)','fontsize',18);


figure(3)
waterfall(t,z_saved,U_saved.*conj(U_saved));
set(gca,'linewidth',2,'fontsize',16,'Ticklength',[0.03
0.03]);
xlabel('Time(ps)','fontsize',18);
ylabel('Distance(m)','fontsize',18);
zlabel('Intensity (arb.units)','fontsize',18)
```

www.ingramcontent.com/pod-product-compliance
Ingram Content Group UK Ltd.
Pitfield, Milton Keynes, MK11 3LW, UK
UKHW051852100926
8556IPUK00005B/592

9 780750 326841